D1629637

Macromolecular Syntheses

1807 175 YEARS OF PUBLISHING 1982

Macromolecular Syntheses

A Periodic Publication

of Methods for the Preparation

of Macromolecules

VOLUME EIGHT   *ELI M. PEARCE, Editor*

1982

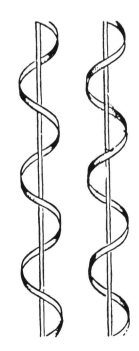

A Wiley-Interscience Publication
John Wiley & Sons
New York • Chichester • Brisbane • Toronto • Singapore

Library of Congress Catalog Card Number: 63-18627

ISBN 0-471-86876-0

Printed in the United States of America

10 9 8 7 6 5 4 3 2 1

Submission of Preparations

All chemists are invited to submit procedures for the preparation of polymers to *Macromolecular Syntheses*. Preparations of both old and new polymers that are of general interest or that illustrate useful preparative techniques are desired. Chemists who plan to submit preparations are urged to write to the secretary concerning their intentions.

The author should carefully check his preparation before submitting it. Explicit directions, including elaboration in notes where necessary, for all the steps of the preparation and the isolation of the polymer should be given. The synthesis of the starting materials or their source along with criteria to be used in determining their purity should be given. Experience in checking procedures for preparing polymers has shown that impure or uncharacterized starting materials present the greatest difficulty in duplicating the author's results. The range of yields and polymer property values should be reported rather than the maximum values obtainable. The characterization of the polymer should include such information as molecular weight, viscosity, softening or melting point, spectra, functional group analysis, and solubility data.

Authors should submit three copies of their preparation to the secretary of *Macromolecular Syntheses:*

Professor David A. Tirrell
Department of Chemistry
Carnegie Mellon University
4400 Fifth Avenue
Pittsburgh, Pennsylvania 15213

Preface

This volume appears during a period in which macromolecular synthesis is gaining increasing importance. The 1980's will be a time in which the opportunities for specialty polymers will be maximized. Diverse areas such as oil recovery, computers, communications, engineering applications, food, and health will require and utilize new polymers. *Macromolecular Syntheses* is an important contribution to the progress of the specialized polymer area because it is a repository of polymer synthesis and its procedures, which have been carefully described and tested in sufficient detail so that reproducible syntheses can be made.

Many thanks are due to the individuals who contributed to this and the previous volumes. We encourage investigators to submit new preparations so that this successful series will continue to grow. Special thanks must go to the checkers of the procedures, who, by their diligence, have shown a devotion to their profession and this series at the highest level.

I must acknowledge the valiant efforts of Professor James A. Moore and the Editorial Board for the privilege of working with them in the production of this volume. I thank Dr. Robert B. Fox for supplying the structure-based names [see *Pure and Applied Chemistry* **48**, 373(1976)] included on the first page of most of the preparations.

ELI M. PEARCE

Brooklyn, New York
July 1982

Contents

Macromolecular Syntheses

N-Chloronylon 66

{Poly[(Chloroimino)adipoyl(chloroimino)-hexamethylene]}

$$+\text{NH}-(\text{CH}_2)_6-\text{NH}-\text{CO}-(\text{CH}_2)_4-\text{CO}+_n \xrightarrow{\textit{tert}\text{-BuOCl}}$$

$$+\text{N}-(\text{CH}_2)_6-\text{N}-\text{CO}-(\text{CH}_2)_4-\text{CO}+_n$$
$$\overset{|}{\text{Cl}}\phantom{-(\text{CH}_2)_6-}\overset{|}{\text{Cl}}$$

Submitted by H. Schuttenberg, K. Hahn, and R. C. Schulz (1)
Checked by M. Parpart, R. W. Campbell, and O. Vogl (2)

Nylon 66 can be converted with *tert*-butyl hypochlorite under mild conditions into its N-chloro derivative. The chlorination proceeds in high yields and side reactions occur only to a slight extent.

1. Procedure

Commercial nylon 66 must be purified by reprecipitation prior to its use for chlorination.

A. Reprecipitation of Nylon 66

Commercial nylon 66 (40 g) (Note 1) is pulverized and dissolved in 200 ml of concentrated formic acid (98-100%), and then the solution, which sometimes has a faint brownish color, is poured into 1 liter of methanol. The viscous solution is allowed to stand for 1 hr, 1 liter of distilled water is added with stirring, the precipitated polyamide is filtered after 1½ hr through a filter funnel, and the filter cake is washed several times with hot water. In a 500-ml two-necked flask with a mechanical stirrer, the damp powder is suspended in 200 ml of toluene and the water is removed by azeotropic distillation (with stirring) and is collected in a Dean-Stark trap. The suspension is now filtered through a pressure filter, washed with anhydrous ether, and dried over P_2O_5. Nylon 66 is obtained in a finely divided form in a yield of 36 g (90% of theory) (Note 2).

B.   Chlorination with tert-Butyl Hypochlorite

In a 50-ml Erlenmeyer flask, reprecipitated nylon 66 (1.13 g, 0.005 mole, based on nylon 66 repeat unit) (Note 3) is swollen in 15 ml of anhydrous 1,1,2,2-tetrachloroethane (Note 4). The reaction mixture is stirred for 1 hr in the closed flask. Under exclusion of light at 10°, 0.05 ml of concentrated formic acid and 2.1 ml of tert-butyl hypochlorite (0.02 mole) is added (Note 5). The mixture is stirred for 2-4 hr and a clear solution of N-chloronylon 66 is obtained. A small amount of undissolved, only partially chlorinated, nylon 66 is some-times encountered. The solution is allowed to drop quickly into 300 ml of anhydrous ether with stirring (Note 6). Precipitated N-chlorinated nylon 66 is collected after 2-3 hr of stirring by pressure filtration through a coarse filter funnel. The solid is washed several times with anhydrous ether and dried (Note 7). The product is dissolved in 50 ml of 1,1,2,2-tetrachloroethane or chloroform (at temperatures not exceeding 50°) and the polymer is precipitated by pouring the solution into 150 ml of anhydrous ether. If the solution is not clear prior to precipitation (because of small amounts of only partially chlorinated nylon 66), filtration or centrifugation may be necessary. After filtration, washing, and drying under reduced pressure, a colorless, fluffy powder is obtained which should be stored in the absence of light and moisture and at temperatures not above room temperature (Note 8). The yield is 1.25 g (85% of theory).

2. Characterization

N-chloronylon 66 is readily soluble in chlorinated hydrocarbons, benzene, formic acid, and trifluoroacetic acid. The content of active chlorine is determined iodometrically. N-chloronylon 66 (50-10 mg) is dissolved in 20 ml of concentrated formic acid, sodium acetate ($CH_3COONa \cdot 3H_2O$) (5 g) is added, and the solution is treated with KI (0.5 g) in 0.5 ml of water. Iodine is liberated and, after exactly 10 min, the iodine is titrated with 0.1 N sodium thiosulfate solution (Note 9). The blank value of the reagents is determined separately with thiosulfate solution. The content of active chlorine in several preparations of N-chloronylon 66 is 22-23.5 wt.%. This value corresponds to a conversion of 91-97% of theory. The formation of an N-chloroamide group is noticeable in the infrared spectrum by the essential disappearance of the N—H stretching frequency and the amide II band. The amide carbonyl frequency is also shifted by about 35 cm^{-1} to higher wave numbers. The solution viscosity η_{spec}/c in chloroform at 25° (Ostwald viscometer, capillary diameter 0.32 mm) is 0.63-0.72 dl/g at a concentration of $c = 1$ g/100 ml. When the product is stored at $-15°$ for several months, the content of active chlorine drops only slightly.

3. Notes

1. Nylon 66 must be precipitated and carefully dried prior to chlorination, otherwise the chlorination proceeds slowly and incompletely (3,4).

2. The molecular weight of the precipitated nylon 66, calculated from the intrinsic viscosity (5,6), is $M_v = 27,700$.

3. In this preparation only the chlorination of a 1-g quantity of nylon 66 is described; however, a multiple of this amount can be chlorinated without difficulties.

4. Technical 1,1,2,2-tetrachloroethane is washed repeatedly with concentrated sulfuric acid followed by washings with water and saturated potassium carbonate solution, until the solution is colorless. After drying with anhydrous potassium carbonate for one week, 1,1,2,2-tetrachloroethane is rectified under reduced pressure. Ether and toluene are purified and dried according to standard methods.

5. *Tert*-butyl hypochlorite can be prepared by a method described in *Organic Syntheses* (7).

6. All solvents must be rigorously dried and all preparations must be carried out under complete exclusion of moisture to obtain good results.

7. Should *N*-chloronylon 66 precipitate in a finely divided form which is difficult to filter, it is preferrable to decant the main portion of the supernatant liquid and isolate the polymer by centrifugation.

8. Chlorination, isolation, and reprecipitation should be carried out without interruption.

9. During, or after, the titration the formation of a precipitate may be observed. This occurrence does not influence the accuracy of the determination.

4. Other Methods of Preparation of *N*-Chloronylon 66

The chlorination of nylon 66 can also be carried out with alkali metal hypochlorites, hypochlorous acid (HOCl in H_2O), or with dichloromonoxide solution (3,4). Chlorination of nylon 66, especially with dichloromonoxide solutions or elemental chlorine, causes a relatively large decrease in the degree of polymerization; with *tert*-butyl hypochlorite the molecular weight decrease is substantially smaller.

Many other polyamides of the **AABB** type or **AB** type, as well as other unsubstituted polymers, can be chlorinated in a similar manner. The maximum degree of chlorination, side reactions, degradation, and the stability of the resulting polymers differ substantially and depend very much on the type of the original polymer.

5. References

1. Institute of Organic Chemistry, University of Mainz, D-6500 Mainz, Germany.
2. Polymer Science and Engineering Department, University of Massachusetts, Amherst, Massachusetts 01002.
3. H. Schuttenberg and R. C. Schulz, *Makromol. Chem.* **143**, 153 (1971).
4. H. Schuttenberg and R. C. Schulz, *Angew. Makromol. Chem.* **18**, 175 (1971).
5. D. Braun, H. Cherdron, and W. Kern, *Praktikum der Makromolekulare Organische Chemie*, A. Hüthig-Verlag, Heidelberg, 2nd Edition, 1971; *Techniques of Polymer Synthesis and Characterization*, Wiley-Interscience, New York, 1972, p. 69.
6. H. G. Elias and R. Schumacher, *Makromol. Chem.* **76**, 23 (1964).
7. N. Rabjohn (editor), *Organic Syntheses, Collective Vol. 4*, Wiley, New York, 1966, p. 125.

Poly(Carbamoyl Disulfide) Based on 1,4-Butanedithiol and 1,3-Di-4-Piperidylpropane

{Poly(1,4-Piperidinediyltrimethylene-4,1-Piperidinediylcarbonyldithiotetramethylene-dithiocarbonyl)}

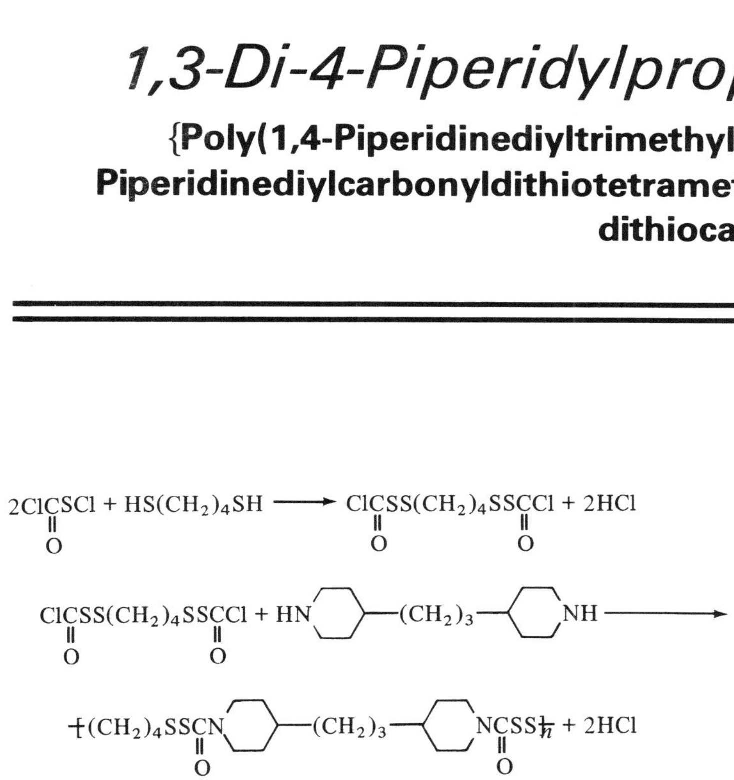

Submitted by N. Kobayashi, A. Osawa, and T. Fujisawa (1)

Checked by R. F. Deveux and E. J. Goethals (2)

1. Procedure

A. 1,4-Butane-bis(Dithiocarbonyl Chloride)

Caution! The operation should be carried out in a well-ventilated hood. A 100-ml three-necked flask is equipped with a stirrer, a thermometer, a dropping funnel, and a drying tube. In the flask are placed 26.0 g (0.198 mole) of freshly distilled chlorocarbonylsulfenyl chloride (Note 1) and 30 ml of *n*-hexane (Note 2). The flask and its contents are cooled to $-10°$. A solution of 12.0 g (0.098 mole) of 1,4-butanedithiol (Note 3) in 10 ml of *n*-hexane is added dropwise for 15 min to the stirred reaction mixture (Note 4). The residual dithiol is washed into the flask with another 5-ml portion of *n*-hexane. The resulting heterogeneous solution is warmed slowly to room temperature and stirred at room temperature until no hydrogen chloride is evolved (Note 5 and 6). The *n*-hexane and excess chlorocarbonylsulfenyl chloride are removed thoroughly under reduced pressure to give a yellowish liquid with an unpleasant odor. The yield is 28.9-30.1 g (95-99%) (Notes 7 and 8).

B. Polycondensation

A mixture of 2.67 g (12.7 mmole) of 1,3-di-4-piperidylpropane (Note 9), 1.02 g (25.4 mmole) of sodium hydroxide, 20 ml of chloroform (Note 10), and 400 ml of cold water is stirred rapidly in a high-speed blender (Note 11). To this mixture is quickly added a solution of 3.95 g (12.7 mmoles) of 1,4-butane-*bis*(dithiocarbonyl chloride) in 60 ml of chloroform, and the resulting mixture is stirred for 5 min at a high speed. The chloroform is removed using a rotary evaporator. The solid product is collected by suction filtration, washed several times with water, then with methanol (Note 12), and finally dried at $50°$ in a vacuum (Note 13). The yield is nearly quantitative. The polymer is purified by reprecipitation from chloroform into methanol (Note 14).

2. Characterization

Analysis. Calcd. for $(C_{19}H_{32}N_2O_2S_4)_n$: C, 50.86%; H, 7.19%; N, 6.24%; S, 28.58%. Found: C, 50.87%; H, 7.39%; N, 6.39%; S, 28.72%.

The infrared spectrum of the product shows a characteristic band at 1670 cm^{-1} for carbonyl groups. The polymer is soluble in chloroform, *sym*-tetrachloroethane, and *m*-cresol, and is insoluble in *N,N*-dimethylformamide and hexamethylphosphorictriamide. The inherent viscosity of the polymer is 1.86 dl/g, determined in chloroform at $30°$ for a concentration of 0.5 g in 100 ml of solvent (Note 15). Transparent, tough films can be prepared by drying chloroform solutions. The polymer films decompose upon ultraviolet irradiation with liberation of carbonyl sulfide (Note 16). The polymer is crystalline as determined by x-ray and has a melting point of $142-147°$ measured on a Yanaco MP micro-melting-point apparatus.

3. Notes

1. Chlorocarbonylsulfenyl chloride is prepared from trichloromethanesulfenyl chloride by partial hydrolysis in concentrated sulfuric acid (3,4) and is commercially available from Fluka.

2. *n*-Hexane is distilled and dried over sodium wire.

3. 1,4-Butanedithiol is purified by distillation; b.p. 89-90° (22 torr).

4. Because the reaction is very exothermic the dithiol should be added slowly to prevent the temperature of the reaction mixture from exceeding 0°.

5. About 1 hr is required for completion of the reaction.

6. In some cases a small amount of solid material formed on the wall of the flask. It can be removed by suction filtration.

7. Because the 1,4-butane *bis*(dithiocarbonyl chloride) is analytically and spectrally homogeneous, it can be used for the subsequent polycondensation without further purification. The infrared spectrum shows $\nu_{C=O}$ at 1770 cm^{-1} and ν_{C-Cl} at 790 cm^{-1}; nmr (CDCl$_3$) shows δ 1.84 (m, 4H) and 2.93 (m, 4H). Attempts to distill the product were unsuccessful because of its poor thermal stability.

8. Storage of this reagent in a deep freeze is recommended to avoid formation of the brown color that develops after 24 hr at room temperature.

9. Material from Tokyo Chemical Industry Co., 9-4 Nihonbashi-Honcho 3-chome, Chuo-Ku, Tokyo, Japan was satisfactory without further purification.

10. Chloroform is freed of the alcohol present as a stabilizer by washing five times with equal volumes of water.

11. High-speed stirring is needed to disperse the reactants rapidly. The checkers successfully used a round-bottomed flask and a laboratory stirrer operating at 4000 rpm.

12. Methanol helps to remove the water-immiscible solvent and increases the drying rate. Stirring in the blender will speed the washing.

13. The checkers obtained a rubbery mass which was washed with water and methanol by decantation rather than filtration.

14. The checkers dissolved the product in 60 ml of chloroform and precipitated with 1500 ml of methanol. If a more dilute solution in chloroform was used, the precipitated polymer was a fine suspension which was difficult to separate by filtration.

15. The checkers were unable to prepare polymer with a comparable inherent viscosity and obtained instead 0.98 dl/g. They found, moreover, that the viscosity decreased to 0.93 in only 15 min. This apparent degradation was observed in the dark as well as in the light. The polymer of the submitters did not show this behavior.

16. This is a common feature for polymers containing $-XSC(=O)Y-$ (X, Y = S or N) linkages in the main chain (3,5,6,7).

4. Methods of Preparation

No other methods of preparation for this particular polymer have been reported. This method has been used for similar poly(carbamoyl disulfides) (7).

5. References

1. Sagami Chemical Research Center, Kanagawa, Japan.
2. Rijksuniversiteit Gent, B-9000 Gent, Belgium.
3. N. Kobayashi and T. Fujisawa, *Macromolecules* 5, 106 (1972).
4. *Neth. Appl.* 6,514,548; *Chem. Abstr.* 65, 12112h (1966).
5. N. Kobayashi and T. Fujisawa, *J. Polym. Sci. A-1* 10, 1233 (1972).
6. N. Kobayashi and T. Fujisawa, *J. Polym. Sci. Polym. Chem. Ed.* 10, 3165 (1972).
7. N. Kobayashi, A. Osawa, and T. Fujisawa, *J. Polym. Sci. Polym. Chem. Ed.* 11, 553 (1973).

Poly(Hexamethylene trans-4-Octen-1,8-Diamide by Interfacial Polycondensation

[Poly(Iminohexamethyleneiminocarbonyl-*trans*-3-Hexenylene)]

$$\text{ClOC}\underset{\text{CH}_2}{\overset{\text{CH}_2}{\diagdown}}\underset{\text{CH}}{\overset{\text{CH}}{=\!\!=}}\underset{\text{CH}_2}{\overset{\text{CH}_2}{\diagup}}\text{COCl} + \text{H}_2\text{N}-(\text{CH}_2)_6-\text{NH}_2$$

$$\downarrow \text{NaOH}$$

$$+\!\!\text{CO}-(\text{CH}_2)_2-\text{CH}=\text{CH}-(\text{CH}_2)_2-\text{CO}-\text{NH}-(\text{CH}_2)_6-\text{NH}\!\!+_{\!\overline{n}}$$

Submitted by G. Maglio and R. Palumbo (1)

Checked by P. Maravigna and G. Montaudo (2)

1. Procedure

1,6-Diaminohexane (1.39 g, 0.012 mole) (Note 1) and sodium hydroxide (0.96 g, 0.024 mole) are dissolved in 300 ml of water and placed in a 1-liter high-speed blender (Note 2). The blender is turned to maximum speed and *trans*-4-octen-1,8-dioyl chloride (2.51 g, 0.012 mole) (Note 3) in 60 ml of chloroform-hexane (50/50 by volume) (Note 4) is added as quickly as possible. The stirring is discontinued after 10 min and the polymer is collected by filtration on a medium-pore fritted glass filter.

The polymer is washed repeatedly with 300 ml of distilled water by stirring in the blender (Note 5), collected again on the glass filter, and then washed well with acetone. After drying in a vacuum oven at 80° for 12 hr, a 65-75% yield of the product is obtained.

2. Characterization

The polymers have inherent viscosities of 1.8-2.3 dl/g in *m*-cresol at 25° at a concentration of 0.05 g/dl. The polymer is soluble in formic acid, dichloroacetic acid, and *m*-cresol. Smooth films may be obtained by slow evaporation of formic acid solutions.

The melt temperature of the polymer determined by DSC under nitrogen at 16°/min is 259°. $\bar{M}_n$ determined by end-group titration for a polymer with η_{inh} = 1.89 dl/g in *m*-cresol (0.5 g/dl, 25°) (3) is 28,000 ± 1,000.

Infrared analysis performed on films shows the following absorptions of the amide group: 1635(s), 1545(s), 1213(m), 695(w), and 590(vw) cm^{-1}. The trans double-bond absorption is found as a sharp peak at 980 cm^{-1}. The elemental analysis of a well-dried sample gave: *Analysis.* Calcd. for $(C_{14}H_{24}N_2O_2)_n$: C, 66.66%; H, 9.52%; N, 11.11%. Found: C, 66.24%, H, 9.84%; N, 11.14%.

3. Notes

1. 1,6-Diaminohexane is recrystallized twice from dry toluene and stored under dry nitrogen.

2. The blender jar, previously cooled in a freezer, was covered with aluminium foil and over this is placed the plastic cap. A wide-mouthed funnel is inserted through the foil for addition of the diacyl halide.

3. The *trans*-4-octen-1,8-dioyl chloride was prepared by refluxing *trans*-4-octen-1,8-dioic acid (4) with an excess of thionyl chloride for 4 hr. The crude acid dichloride was distilled twice under reduced pressure through a 10-in. Vigreux column, with b.p. 90° at a pressure of 0.05 torr.

4. Chloroform, reagent grade, is freed from the alcohol present as stabilizer just before use according to the following procedure: 100 ml are washed twice with water, allowed to stand over calcium chloride for 3 hr, and finally passed through a 1 X 7-in. column filled with Merck basic aluminum oxide. The first 15 ml are rejected and the purified chloroform is collected under nitrogen. Reagent grade *n*-hexane is dried by passing through a column filled with Merck neutral aluminum oxide.

5. Several washings are required to remove NaCl.

6. *n*-Octane and carbon tetrachloride may also be used as the organic phase in this procedure (5). The inherent viscosities of the resulting polymer fall in the range 1.5-2.0 dl/g (*m*-cresol, 25°, $c = 0.5$).

4. Methods of Preparation

Polyamides from unsaturated acids such as fumaric or substituted fumaric acids have been prepared by other authors using either interfacial or solution polycondensation (5,6).

5. References

1. Laboratorio di Richerche su Tecnologia dei Polimeri e Reologia del C.N.R., Arco Felice (Napoli) Italy.
2. Universita Di Catania, Viale A. Doria 8, 95125 Catania, Italy.
3. E. Waltz and G. B. Taylor, *Anal. Chem.* **19**, 448 (1947).
4. K. Shishido, K. Sei, and H. Nozaki, *J. Org. Chem.* **27**, 2681 (1962).
5. N. Lanzetta, G. Maglio, C. Marchetta, and R. Palumbo, *J. Polym. Sci. Chem. Ed.* **11**, 913 (1973).
6. V. Guidotti, M. Russo, and L. Mortillaro, *Makromol. Chem.* **147**, 111 (1971).

Polyelectrolytes from 2,3-bis(Bromomethyl)-1,3-Butadiene and Tertiary Diamines

{Poly[(Dimethyliminio)trimethylene(dimethylminio)(2,3-Dimethylenetetramethylene)-Dibromide]}

Submitted by R. M. Ottenbrite and G. R. Myers (1)
Checked by L. K. Post and B. M. Culbertson (2)

1. Procedure

A. *Monomer Synthesis*

A 200-ml stainless steel Parr general-purpose bomb is charged with 41 g (0.5 mole) of 2,3-dimethyl-1,3-butadiene (Note 1), 20 ml of methanol, and 1 g of hydroquinone. The vessel is then cooled in a dry-ice-isopropyl alcohol bath and 32 g (0.5 mole) of liquid sulfur dioxide is added. The bomb is sealed, heated slowly to 85° in a water bath, and maintained at that temperature for 4 hr. It is then cooled, the product is removed, and the bomb is rinsed with methanol. The sulfone and washings are combined and dissolved in hot methanol and treated with 2 g of Norite. The filtrate is concentrated to 200 ml and the sulfone is allowed to crystallize under refrigeration. Recrystallization from methanol yields 60-64 g (82-88%) melting at 135° (Note 2) (recovery from the mother liquor raises the yield to 65-70 g).

A solution of 58.4 g (0.4 mole) of the sulfone in 350 ml of methylene chloride (Note 3) is treated with 135 g (0.76 mole) of *N*-bromosuccinimide in a 1-liter flask fitted with a condenser. The flask is carefully irradiated with a 275-W GE sunlamp (Note 4). Once the reaction begins, the distance between the lamp and the flask must be adjusted to maintain controlled reflux. After the reaction is completed, the flask is cooled to 5° overnight (*Caution!* Note 5) and the precipitated succinimide is removed by filtration. The filtrate is concentrated on a rotary evaporator to a syrup which is then stirred with an equivalent volume of 95% ethanol. Precipitation of the 2,3-*bis*(bromomethyl)-2,5-dihydro-thiophene-1,1-dioxide occurs in 15-30 min. Recrystallization from hot methanol gives 46-54 g (38-45%) of the dibromosulfone melting at 119-122°.

The decomposition of 9.06 g (0.03 mole) of the dibromosulfone is carried out in a 25-ml flask joined by a short connecting tube to a 25-ml receiver, cooled in an ice bath (Note 6) and equipped with a vacuum outlet. A vacuum trap cooled in liquid nitrogen is used to condense liberated sulfur dioxide. The system pressure is brought to 0.5 torr and the reaction flask is heated with an oil bath at 160° which can be slowly increased to a maximum of 170° during the decomposition. (The decomposition is complete in 1-2 hr.) The 2,3-*bis*(bromo-methyl)-1,3-butadiene is recrystallized from hexane to yield 5.0-5.4 g (70-75%) with m.p. 57-58° (Note 7). This monomer is sublimed at 50°/0.1 torr onto a cold finger cooled with a dry-ice-isopropanol slurry and protected from light.

B. Polymerization

A mixture of 1.30 g (0.01 mole) of freshly distilled *N,N,N',N'*-tetramethyl-1,3-propane diamine (Note 8) in 15 ml of methanol is placed in a 50-ml screw cap bottle equipped with a magnetic stirring bar. An equivalent amount of 2,3-*bis*(bromomethyl)-1,3-butadiene [2.40 g (0.01 mole)] is added and the bottle is flushed with nitrogen and sealed. The bottle is kept in a 25° bath and stirred for 100 hr. Precipitation occurs during this time. This mixture is poured into 100 ml of acetone and the polymer is collected by filtration, washed with dry acetone, and dried under vacuum. The polymer yield is 3.3-3.5 g (90-96%).

2. Characterization

The above procedure gives a cationic polyelectrolyte with an intrinsic viscosity of approximately 0.12 dl/g in 0.4 M potassium bromide at 30°, determined with a No. 50 Cannon-Ubbelohde viscometer. A plot of η_{spec}/c vs. C at four polymer concentrations is used to determine the intrinsic viscosity. The polymer is very soluble in water and much less soluble in methanol and dimethylformamide.

3. Notes

1. Commercial 2,3-dimethyl butadiene can be used directly. A freshly distilled sample will give higher yields while less expensive commercial samples of 90-95% purity have been used successfully with a proportionate decrease in yield.

2. Melting points were measured on a Thomas-Hoover melting-point apparatus in sealed capillaries and are uncorrected.

3. Chloroform from which the ethanol preservative has been removed is also an effective solvent (3).

4. An ordinary 100-W tungsten light bulb can also be used if the reaction flask is also heated until reaction begins.

5. All materials from this stage are lachrymators and skin irritants and rubber gloves should be worn when working with them.

6. Both the connecting tube and the receiver should be protected from light to prevent polymerization.

7. This compound should be stored in the absence of light and in the freezer because it readily forms an insoluble polymer.

8. Other N,N,N',N'-tetramethylalkyldiamines have been used with equal success (4) with the exception of methylenediamine, which forms cyclic compounds (5,6).

4. References

1. Department of Chemistry, Virginia Commonwealth University, Richmond, Virginia 23284.
2. Ashland Chemical Co., 5200 Paul G. Blazer Memorial Parkway, Dublin, Ohio 43017.
3. G. B. Butler and R. M. Ottenbrite, *Tetrahedron Lett.*, 4873 (1967).

4. R. M. Ottenbrite and G. R. Myers, *J. Polym. Sci. Chem. Ed.* **11**, 1443 (1973).
5. R. M. Ottenbrite and G. R. Myers, *Can. J. Chem.* **51**, 3631 (1973).
6. H. Noguchi and A. Rembaum. *Macromolecules* **5**, 253 (1972).

Alternating Copolydisulfide of Ethanedithiol and 1,10-Decanedithiol

[Poly(Dithioethylenedithiodecamethylene)]

$$2C_2H_5O\underset{\underset{O}{\|}}{C}SCl + HS(CH_2)_2SH \longrightarrow C_2H_5O\underset{\underset{O}{\|}}{C}SS(CH_2)_2SS\underset{\underset{O}{\|}}{C}OC_2H_5 + 2HCl$$

$$nC_2H_5O\underset{\underset{O}{\|}}{C}SS(CH_2)_2SS\underset{\underset{O}{\|}}{C}OC_2H_5 + nHS(CH_2)_{10}SH \xrightarrow{\text{Et}_3\text{N}}$$

$$\text{+}(CH_2)_2SS(CH_2)_{10}SS\text{+}_{\overline{n}} + 2nCOS + 2nC_2H_5OH$$

Submitted by N. Kobayashi and T. Fujisawa (1)

Checked by R. F. Deveux and E. J. Goethals (2)

1. Procedure

A. 1,2-Bis(ethoxycarbonyldithio)ethane

Caution! These operations should be carried out in a well-ventilated hood. Ethanedithiol, ethoxycarbonyl sulfenyl chloride and gaseous HCl are noxious materials. A 200-ml three-necked flask is equipped with a stirrer, a thermometer, a dropping funnel, and a reflux condenser. In the flask is placed a solution of chloroform (30 ml) (Note 1) containing 8.4 g (89 mmole) of ethanedithiol (Note 2). The mixture is cooled to about $-10°$. A solution of 25.6 g (188 mmole) of ethoxycarbonylsulfenyl chloride (Note 3) in 30 ml of chloroform is added dropwise at temperatures below 0° (Note 4). At the completion of the addition, the cooling bath is removed, the flask is allowed to come to room temperature for 2 hr. The mixture is heated at 55° until no hydrogen chloride is evolved (Note 5). The chloroform is removed using a rotary evaporator. From the residual pale yellow liquid, 1,2-*bis*(ethoxycarbonyldithio)ethane is isolated as white crystals by column chromatography on silica gel (benzene:*n*-hexane = 7:3 by volume). It is purified by recrystallization from *n*-hexane. The yield is 19.0-21.8 g (70-81% based on ethanedithiol). The melting point is 48-49°C (Note 6).

B. Polymerization

A solution of 2.9 g (9.6 mmole) of 1,2-*bis*(ethoxycarbonyldithio)ethane, 2.0 g (9.7 mmole) of 1,10-decanedithiol (Note 7), and 0.1 g (1 mmole) of triethyl-amine in 40 ml of chloroform is stirred for 1 hr at room temperature and then refluxed for 1 hr (Note 8). The reaction mixture is concentrated to 10-15 ml and added to a large, vigorously stirred volume of methanol. The resulting precipitate is collected by suction filtration and dried at 50° in a vacuum. The yield is 95-100%. The polymer is purified by reprecipitation from chloroform into methanol.

2. Characterization

Analysis. Calcd. for $(C_{12}H_{24}S_4)_n$: C, 48.60%; H, 8.16%; S, 43.24%. Found: C, 48.81%; H, 8.29%; S, 43.19%.

The spectrum of the product is devoid of residual C=O absorption. The polymer is soluble in dichloromethane, chloroform, and *sym*-tetrachloroethane. The inherent viscosity of polymer is in the range 0.17-0.34 dl/g at 30° at a concentration of 0.5 g of polymer per 100 ml of *sym*-tetrachloroethane (Note 9). The polymer is highly crystalline as determined by x-ray (Note 10) and has a melting point (Note 11) of 80°, which is higher than that of the corresponding random copolydisulfide (m.p. 43°), prepared from a 1:1 mixture of ethane-dithiol and 1,10-decanedithiol by DMSO oxidation (3). The random copolymer has a low degree of crystallinity.

3. Notes

1. Chloroform was freed of alcohol stabilizer by washing five times with equal volumes of water.

2. Ethanedithiol was purified by distillation; b.p. 146°.

3. Ethoxycarbonylsulfenyl chloride can be prepared by the reaction of

chlorocarbonylsulfenyl chloride (4) with ethanol (5) (Note 12). Methoxy-carbonyl sulfenyl chloride is commercially available from Fluka.

4. Evolution of hydrogen chloride begins in the initial stage of the addition. The reaction is exothermic.

5. It requires from 0.5 to 2 hr for completion of the reaction.

6. The melting point is observed on a hot-stage microscope.

7. 1,10-Decanedithiol is purified by distillation; b.p. 169-170° (16 torr). The checkers used 1,10-decanedithiol supplied by Fluka A. G., Switzerland.

8. The refluxing may be continued for a longer time, but there is little change after 1 hr.

9. The checkers found a value of 0.30 dl/g at 30° in chloroform.

10. X-ray powder pattern [2θ (relative intensity to $2\theta = 20.0°$ as 100%)]: 5.7° (18); 11.8° (12); 19.0° (42); 20.0° (100); 21.7° (36); 22.8° (65); 26.7° (18).

11. The melting points of the polymers were determined by DTA with a AGNE TGD-1500 thermal balance in air at a programmed heating rate of 10°/min. The checkers found a value of 81° by differential scanning calorimetry.

12. Ethyl alcohol [46 g (1 mole)] is added slowly dropwise at 10° to chloro-carbonylsulfenylchloride [131 g (1.0 mole)] with stirring while excluding moisture. The mixture is warmed to 50° and stirred for about 2 hr at this temperature while hydrogen chloride is evolved. After gas is no longer elimi-nated, the reaction mixture is fractionally distilled to yield 111 g (78%) of ethoxycarbonylsulfenyl chloride boiling at 45-47°/15-16 torr.

4. Methods of Preparation

No other methods of preparation for this particular polymer have been described. This method has been used for similar copolydisulfides (6).

5. References

1. Sagami Chemical Research Center, Kanagawa, Japan.
2. Rijksuniversiteit Gent, B-9000 Gent, Belgium.
3. E. J. Goethals and C. Sillis, *Makromol. Chem.* **119**, 249 (1968).
4. N. Kobayashi and T. Fujisawa, *Macromolecules* **5**, 106 (1972).
5. W. Weiss and E. Mühlbauer, Ger. Offen. 1, 568,632 (Farbenfabriken Bayer AG), March 19, 1970.
6. N. Kobayashi and T. Fujisawa, *J. Polym. Sci. Polymer Chem. Ed.* **11**, 545 (1973).

Head-to-Head Poly-(Methyl Cinnamate) from Alternating Stilbene-Maleic Anhydride Copolymer

{Poly[1,2-Bis(Methoxycarbonyl)-3,4-Diphenyl-tetramethylene]}

Submitted by T. Tanaka and O. Vogl (1)
Checked by J. C. Salamone, M. U. Mahmud, and M. Mitry (2)

1. Procedure

A. Preparation of Alternating Stilbene and Maleic Anhydride Copolymer (Note 1)

Maleic anhydride (*Caution! Maleic anhydride is toxic and skin contact should be avoided*) [3.93 g (40 mmole) Eastman] and *trans*-stilbene [7.20 g (40 mmole)] (Note 2) are dissolved in 50 ml of benzene (*Caution! Benzene is toxic and should be used in a hood.*) (Notes 3 and 4) at room temperature. Azobisisobutyronitrile [33 mg (0.20 mmole), 0.50 mol. % based on each monomer] is added, and the solution is charged to a 100-ml polymerization tube to which a three-way stopcock is attached at the open end of the tube. The tube is cooled in a dry-ice-methanol bath and degassed at 0.1 torr. The stopcock is closed and the tube is disconnected from the vacuum line and allowed to warm gradually to room temperature. The degassing procedure is repeated three times. Finally the tube is sealed while the lower part of the tube is held in a dry-ice-methanol bath.

The sealed tube, protected with a wire sleeve, is held in a constant temperature bath at $60 \pm 2°$.

After 2 hr (Note 5), the tube is cooled in an ice-water bath and opened. The contents of the tube, consisting of a gelatinous mass of swollen copolymer dispersed in benzene, is poured into 500 ml of dichloroethane to coagulate the polymer. The material is washed by decantation three times with 50 ml of dichloroethane, filtered through a coarse fritted filter, and dried at $60°$ at 1 torr, yielding 2.5 g (23% of theoretical yield based on both monomers charged).

According to carbon/hydrogen analysis, the copolymer has the composition of a 1:1 copolymer of stilbene and maleic anhydride as reported in the earlier literature (3,4).

B. *Preparation of Head-to-Head Poly(Methyl Cinnamate)*

The alternating stilbene-maleic anhydride copolymer (2.5 g) is suspended in 330 ml of aqueous 3 N NaOH solution in a 1.5-liter Erlenmeyer flask and agitated at about $90°$ for 1 hr (Note 6). The suspension is then diluted to about 1 liter by adding distilled water at room temperature. A clear solution of H-H poly(sodium cinnamate) in excess sodium hydroxide is obtained (Note 6) and the solution is cooled to room temperature. Freshly distilled dimethyl sulfate (200 ml) (*Caution!* Note 7), a large excess, is added to the solution with vigorous mechanical stirring. The level of the stirring blade and the stirring rate is adjusted to cause efficient mixing of the two layers. Stirring is continued for 1 hr. As the reaction proceeds the esterified polymer is transferred from the upper (aqueous) layer into the lower (dimethyl sulfate) layer.

The dimethyl sulfate solution (100 ml) containing the esterified polymer is separated from the aqueous layer in a separatory funnel. One liter of methanol is added to the dimethyl sulfate solution with stirring. The resulting suspension is filtered through a coarse fritted filter, washed several times with methanol, and dried at room temperature at 0.1 torr, yielding 2.1 g (73% of theory) (Note 8). The analysis (Note 9), spectral characterization, and thermal degradation behavior indicate that the polymer is pure H-H poly(methyl cinnamate) (5).

H-H poly(methyl cinnamate) is a brittle polymer that is soluble in pyridine, dimethyl formamide, and trifluoroacetic acid, among others. It has no glass transition temperature (T_g) observable by DSC measurements and decomposes above 350° to methyl cinnamate and to a small extent to stilbene and a mixture of dimethyl maleate and dimethyl fumarate.

2. Notes

1. *trans*-Stilbene and maleic anhydride are used as starting comonomers. The stereochemistry of H-H poly(methyl cinnamate) is not known and should not be implied from the way the formulas are written.

2. *trans*-Stilbene (Eastman Chemical Co.) is purified from ethanol by one recrystallization before use. Maleic anhydride (Eastman Chemical Co.), as received, contains small amounts of maleic acid which can be removed by dissolving maleic anhydride (10 g) in benzene (100 ml), filtering, and using the filtrate directly for copolymerization.

3. Instead of benzene, toluene or dry chloroform may be used as a solvent for the stilbene-maleic anhydride copolymerization.

4. Benzene is dried over anhydrous calcium chloride and distilled from sodium metal prior to its use. Chloroform is washed three times with water, dried over anhydrous calcium chloride, and distilled from phosphorous pentoxide. It is also possible to use xylene as a solvent.

5. Longer reaction time increases the conversion of comonomers but not the molecular weight of the copolymer (6).

6. This step transforms the anhydride polymer into an aqueous solution of the sodium salt of H-H poly(cinnamic acid).

7. Dimethyl sulfate is very toxic and all operations involving dimethyl sulfate must be carried out under a well-ventilated hood.

8. H-H poly(methyl cinnamate) prepared in this way has an η_{spec}/c of approximately 0.9 dl/g measured at 30° as a 0.2 wt.% solution in trifluoroacetic acid. The molecular weight of H-H poly(methyl cinnamate) is affected by the ratio of initiator (AIBN) to monomer concentrations or by the reaction temperature but not by the reaction time.

9. Analysis of unesterified carboxyl groups indicates that this H-H poly(methyl cinnamate) contains less than 0.2 mol.% of free carboxyl groups.

3. References

1. Department of Polymer Science and Engineering, University of Massachusetts, Amherst, Massachussetts 01002.
2. Department of Chemistry, University of Lowell, Lowell, Massachusetts 01854.
3. J. Wagner-Jauregg, *Chem. Ber.* **63**, 3212 (1930).
4. F. M. Lewis and F. R. Mayo, *J. Amer. Chem. Soc.* **70**, 1533 (1948).
5. T. Tanaka and O. Vogl, *Polymer J.* **6**, 522 (1974).

α,β-Poly(2-Hydroxyethyl)-DL-Aspartamide

α-linked monomer   β-linked monomer

Submitted by P. Neri and G. Antoni (1)

Checked by M. K. Akkapeddi and H. K. Reimschuessel (2)

1. Procedure

A. Poly-DL-Succinimide

Finely powdered DL-aspartic acid (50 g) and 25 g of 85% orthophosphoric acid are thoroughly mixed in a 2-liter round-bottomed flask, using a long stainless steel spatula (Note 1). The flask is connected to a rotary evaporator fitted with a manometer and heated at 180° for 2.5 hr in an oil bath, at a pressure less than 20 torr (water aspirator) (Note 2). *N,N*-Dimethylformamide (200 ml) is then added to the still hot, viscous, glassy mass, and the flask is left at room temperature and shaken occasionally until the product dissolves completely affording a homogeneous, light brown solution (Note 3). The solution is poured, with constant stirring, into a beaker containing about 1 liter of water and the flaky white precipitate obtained is collected on a sintered glass filter. The flask is rinsed with water to recover all the product. The precipitate is filtered by suction, thoroughly washed several times with water until the filtrate is neutral to pH paper, and dried at 110° for 24 hr. Yield = 33-35 g (90-95%). η_{red} = 0.35-0.42 dl/g (c = 0.5 g in 100 ml of *N, N*-dimethylformamide (Note 4).

B. α,β-Poly(2-Hydroxyethyl)-DL-Aspartamide (PHEA)

Poly-DL-succinimide (30 g) is dissolved in 150 ml of *N,N*-dimethylformamide in a 250-ml three-necked flask fitted with a stirrer and a thermometer (Note 5). Ethanolamine (45 ml) is added drop by drop, over a period of about 15 min, with constant stirring. The reaction temperature is maintained at 25-30° with an ice-water bath (Note 6). The solution is stirred for 2 hr at room temperature and excess ethanolamine is neutralized by the dropwise addition of 30 ml of glacial acetic acid, at 30-40°, with external cooling. The solution is diluted with about 300 ml of water, dialyzed (Note 7) for 3 days against running tap water and two days against several changes of distilled water, and then lyophyllized. Yield = 38-42 g (80-86%), η_{red} = 0.23-0.27 dl/g (c = 0.5% in H_2O) (Note 8). PHEA can also be isolated by precipitation in a nonsolvent as follows. At the end of the reaction between polysuccinimide and ethanolamine (without addition of acetic acid) the solution (about 200 ml) is diluted with 200 ml of *N,N*-dimethylformamide and slowly poured, with constant stirring, into a flask containing 500 ml of *n*-butanol. The precipitate is collected on a large sintered glass filter and washed several times with acetone, until the filtrate, diluted with water, is neutral to pH paper. The precipitate is dried in a vacuum at room temperature. The filtration of PHEA from the precipitation mixture is quite slow at the beginning. After most of the mother liquor has been removed, it is best to change the filter to facilitate washing. The original mother liquor is milky because of the presence of colloidal PHEA; however, the loss of polymer is minimal. In a pilot experiment, starting with 3 g of polysuccinimide (reduced viscosity = 0.42 dl/g) the yield of PHEA was about 3.7 g with a reduced viscosity of 0.25 dl/g.

2. Characterization (7)

Sedimentation velocity measurements were performed in a Spinco Model E ultracentrifuge, using a synthetic boundary cell and the phase-plate Schlieren optical system. The extrapolated value of $S^{o}_{20,w}$ was 2.59 Svedburg units. Sedimentation equilibrium measurements were also carried out in the Spinco Model E centrifuge. Viscosity measurements were made in water at 25° with Ostwald viscometers. The relationship between intrinsic viscosity $[\eta]$ and weight-average molecular weight M_w is

$$[\eta] = 2.32 \times 10^{-5} M_w^{0.87}$$

The ir spectrum of PHEA shows strong bands at 3300 cm^{-1} (O-H stretching), 1645 cm^{-1} (amide 1), and 1540 cm^{-1} (amide 2).

3. Notes

1. The polymerization of 50 g of aspartic acid requires a flask of at least 2-liter capacity. The exact amount of orthophosphoric acid indicated must be weighed into the flask. The components must be well mixed. The preparation of the mixture in the flask is difficult because of the thickness of the paste which forms. A few milliliters of water can be added to the mixture, care being taken to avoid splashing during the first stage of heating. When several polymerizations are to be carried out, the two components can be mixed in a beaker with a mechanical stirrer and polymerized in portions. The polymerization can also be done in a vacuum oven. The paste must be spread on a tray in as thin a layer as possible and polymerized *in vacuo* at 180°. The use of a tray covered with a Teflon® cloth is particularly advantageous because of its inertness to phosphoric acid and the facility with which the polymer is detached after polymerization.

2. The flask must be brought to a temperature of 180° before the vacuum is applied, otherwise a polymer of low reduced viscosity is obtained. The heating bath must have a temperature between 170 and 190° throughout the reaction. The bath should be at 200° initially.

3. The flask must be protected from atmospheric moisture during dissolution, or a partial precipitation of the dissolved polysuccinimide may occur. Complete dissolution requires 1-2 days.

4. The reduced viscosity of the poly-DL-succinimide can be increased by treatment with N,N'-dicyclohexylcarbodiimide. Poly-DL-succinimide (10 g) is dissolved in N,N-dimethylformamide (50 ml); N,N'-dicyclohexylcarbodiimide (0.5 g) is then added and the solution is stirred at room temperature for 12 hr. A small amount of dicyclohexylurea may precipitate and is filtered by suction.

The polymer is precipitated by pouring the solution into water with constant stirring. The polymer is recovered by filtration, washed several times with ethanol, and dried in an oven. Reduced viscosity = 0.70-0.80 dl/g (c = 0.5 g in 100 ml of N,N-dimethylformamide).

5. The preparation of larger amounts of PHEA does not pose problems. The submitters have prepared the polymer in lots of about 1 kg.

6. The temperature must be kept low to reduce the possibility of aminolytic degradation of the polymer. The chosen range of temperature gives the best results. At a lower temperature the reaction time is considerably lengthened without a significant improvement in the molecular weight of the product.

7. Dialysis is performed with commercial tubing (Visking regenerated cellulose) without pretreatment. Ultrafiltration with apparatus from De Dansker Sukkerfabrikker (Denmark) using cellulose acetate membrane (type 800) is also effective.

8. PHEA of reduced viscosity (0.35-0.45 dl/g) can be obtained using polysuccinimide treated with N,N'-dicyclohexylcarbodiimide. In this case, a more dilute solution of polysuccinimide in dimethylformamide must be used, otherwise a very viscous reaction mixture is obtained which is difficult to stir (about 10 ml of N,N-dimethylformamide per gram of polysuccinimide is required).

4. Methods of Preparation

Low-molecular-weight poly-DL-succinimide (anhydropolyaspartic acid) has been prepared by thermal polymerization of aspartic acid in the presence of phosphoric acid at atmospheric pressure (3), or without phosphoric acid under vacuum or by azeotropic removal of water (4).

Polymerization in the presence of a suitable amount of phosphoric acid, in a thin layer under vacuum, gives a polysuccinimide of higher reduced viscosity. The reaction of polysuccinimide with various amines, including ethanolamine, has been reported previously (5). In the present method (6) the conditions for the reaction of polysuccinimide with ethanolamine have been adjusted to give better yields and the highest possible molecular weight.

5. References

1. I.S.V.T. Sclavo, Research Center, 53100 Siena, Italy.
2. Allied Chemical Co., P.O. Box 1021R, Morristown, New Jersey 07960.
3. S. W. Fox and K. Harada, in *A Laboratory Manual of Analytical Methods of Protein Chemistry, Including Polypeptides*, P. Alexander and H. P. Lundgren, Eds., Pergamon Press, Elmsford, New York, 1966, p. 127.

4. J. Kovacs, H. N. Kovacs, I. Konives, J. Csaszar, T. Vajda, and H. Mix, *J. Org. Chem.* **26**, 1084 (1961).
5. H. N. Kovacs, J. Kovacs, M. A. Pisano, and B. A. Shidlovsky, *J. Med. Chem.* **10**, 904 (1967).
6. P. Neri, G. Antoni, F. Benvenuti, F. Cocola, and G. Gazzei, *J. Med. Chem.* **16**, 893 (1973); Ger. Offen. Pat. 2,032,470 (1971) [*Chem. Abstr.* **74**, 100414j (1971)].
7. G. Antoni, P. Neri, T. G. Pedersen, and M. Ottesen, *Biopolymers* **13**, 1721 (1971).

Block Copolymer of Poly(Ethylene Adipate) and the Polycarbonate of Phenolphthalein

{Poly[Oxyethyleneoxy—Poly{[Poly(Adipoyloxy-ethyleneoxy)—Carbonyloxy]—Poly(Oxy-1,4-Phenylenephthalidylidene-1,4-Phenyleneoxy-carbonyl)}—Oxycarbonylimino-1,4-(2-Methyl-phenylene)iminocarbonyl]}

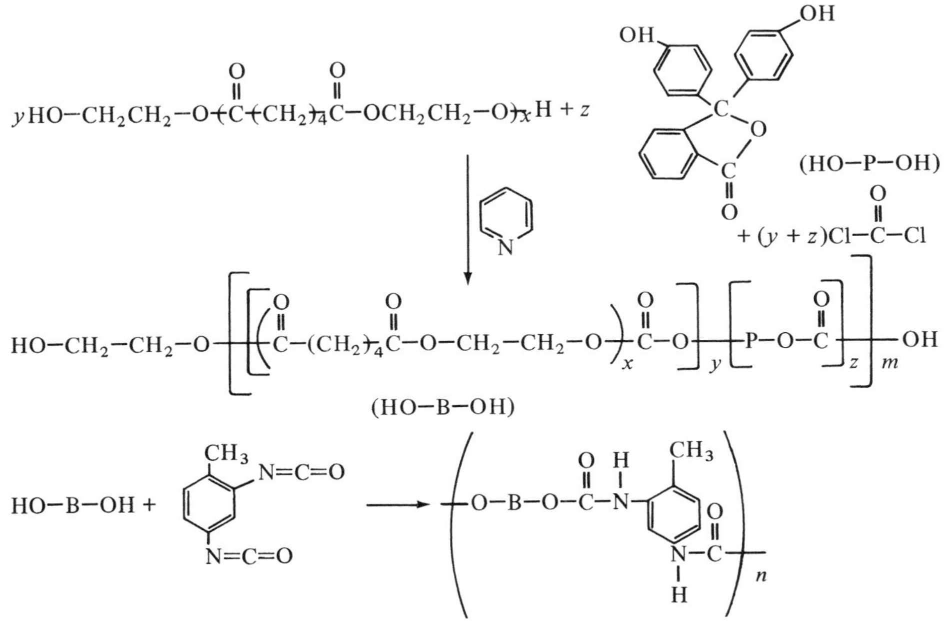

$$y\,HO-CH_2CH_2-O\!\!\left(\!C\!\left(CH_2\right)_4\!C-OCH_2CH_2-O\right)_x H + z$$

(structures)

$$+ (y+z)Cl-C-Cl$$

$$(HO-P-OH)$$

$$HO-CH_2-CH_2-O\left[\left(C-(CH_2)_4C-O-CH_2-CH_2-O\right)_x C-O\right]_y \left[P-O-C\right]_z \Bigg]_m OH$$

$$(HO-B-OH)$$

$$HO-B-OH + \quad \text{(2-methyl phenylene diisocyanate)} \quad \longrightarrow \quad \left(O-B-O-C-N\text{---}N-C\right)_n$$

Submitted by N. A. Memon (1) and H. Leverne Williams (2)

Checked by M. A. Sandhu (3)

1. Procedure

A. *Formation of Multiblock Copolymer*

The method of Perry, Jackson, and Caldwell (4) was modified slightly. A 500-ml three-necked standard taper flask with a motor-driven stirrer, a thermometer, and a calibrated dropping funnel is used. The equipment is carefully cleaned, dried, then assembled, and the flask is immersed in a constant temperature water bath.

To a solution of 13 g of hydroxyl-terminated poly(ethylene adipate) (Multrathane R14 from Mobay) in 100 ml of methylene chloride (*Caution!* Note 1) is added 7 g of phenolphthalein. Upon the addition of 20 ml of pyridine (*Caution!* Note 1) to the stirred suspension the phenolphthalein dissolves to yield a clear solution. Dry reactants and solvents must be used. The temperature of the contents is brought to 30° and maintained there.

The amount of phosgene (*Caution!* Note 2) equivalent to the hydroxyl content of the reactants is calculated and 75% of the molar equivalent is added rapidly (Note 3) as a 12.5% solution in benzene (*Caution!* Note 1). The remaining 25% of the molar equivalent of phosgene is added at a slower rate and a 25% molar excess is added at a still slower rate.

A polymer of acceptable viscosity (0.65 dl/g, measured in chloroform at a concentration of 0.25 g/100 ml at 25°) was also obtained by just maintaining the rate of addition of phosgene in benzene to the stirred reaction mixture so that the mixture was under gentle reflux by the heat of the reaction.

A milliammeter attachment to the stirrer motor was used to note the power consumption, that is, the viscosity of the mixture. When the viscosity increases rapidly, the reaction is stopped. If the reaction mixture is too viscous it is diluted with methylene chloride.

The reaction mixture is washed with water, 0.1 N hydrochloric acid, and again with water to remove the pyridine hydrochloride. One gram of antioxidant (CAO-1, Ashland) (Note 4) is added to the organic phase. The multiblock copolymer is precipitated from the methylene chloride solution by adding an equal volume of acetone first and then the same volume of methanol (or less if precipitation appears to be complete). The precipitate is separated by decantation, shredded by hand, and dried overnight in an oven at 60° and 30 torr.

B. *Chain Extension of the Block Copolymer*

To increase the molecular weight (and possibly to bring about some branching or cross-linking) the block copolymer is redissolved in methylene chloride and an amount of toluene 2,4-diisocyanate (*Caution!* Note 5) equivalent to the calculated hydroxyl content is added. The mixture is stirred thoroughly and poured into poly(tetrafluoroethylene)-lined baking tins. Much of the solvent is evaporated in a fume hood with slow agitation of the solution. When the solution thickens it is placed in a vacuum oven at 60-65°, vacuum is applied to 30 torr carefully to avoid bubbling, and the samples are dried overnight. The films release easily from the pans and can be cut into strips for testing.

2. Notes

1. The solvents are toxic when inhaled and may be irritating on contact. Use a fume hood and avoid contact.

2. Phosgene is a very toxic gas and should not be inhaled. Excess phosgene should be destroyed by bubbling it through water or dilute caustic solution.

3. An excess of phosgene should not be present too quickly because this may result in lower average molecular weight. However, the initial portions may be added quickly and the final portions more slowly to force the reaction to completion. The average equivalent weight of the poly(ethylene adipate) is 1000 g, the equivalent weight of phenolphthalein is 159 g, and the equivalent weight of phosgene is 49.5 g. Thus, in this case, approximately 28 ml of a 12.5% solution of phosgene is required: 17 ml added at an average rate of 2 ml/min, 6 ml at an average rate of 1 ml/min, and 5 ml at an average rate of 0.2 ml/min, that is, dropwise. Note that a solution open to the atmosphere loses phosgene and more of the solution will be required. It is better to use more, or a fresh solution, than to try to concentrate the phosgene. When alicyclic diols are used the methylene chloride is replaced by toluene and the mixture containing the pyridine is refluxed for 5-10 min before the phosgene solution is added: two-thirds of the molar equivalent at a rate of about 1 ml/min at 45°, one-third at a rate of about 1 ml/min at 75°, and the 25% molar equivalent excess at a slower rate of 0.2 ml/min at 85°. The final product in this case is precipitated by pouring into about two volumes of methanol.

4. This treatment inhibits or prevents oxidation of the product on subsequent handling. An excess is used because some is lost in the handling and washing. The specific compound used here is 2,6-di-*t*-butyl-*p*-cresol but other antioxidants can be used.

5. Isocyanates are toxic and should not be inhaled or allowed to come into contact with the skin. Toluene diisocyanate is of sufficiently high boiling point that usual precautions are adequate.

3. Characterization

The intrinsic viscosity was found to be 0.81 dl/g, indicating a fairly high molecular weight which could not be calculated exactly. It would be much higher than twice the average molecular weight of the poly(ethylene adipate), hence the conclusion that the product was multiblock. Nmr analysis of the polymer revealed that the original reactants were in the product in the same ratio as added. Differential scanning calorimetry showed the presence of two blocks having approximately the same (but usually higher) glass transition temperatures than exhibited by the pure homopolymers. There might be an intermediate transition attributable to a "solid solution" (5). Films were analyzed by stress-strain and stress-optical techniques and the results have been published elsewhere (6).

4. Methods of Preparation

This procedure can be used to make a most diverse range of products. Both the elastomeric block and the diol used to make the polycarbonate blocks may be varied. The incorporation of an isocyanate reaction enables further chain extension or the incorporation of other blocks.

The choice of materials would be made on the basis of modulus desired, hydrolytic stability needed, and service conditions such as temperature. The ratio of the two starting materials can be adjusted both ways from the 65/35 ratio used here. More polycarbonate makes the polymer harder with higher modulus and stiffness. More elastomeric block results in lower hardness and higher elasticity. However, a comparatively small increase in the ratio of the elastomeric portion to 70/30 may yield a viscous product rather than an elastomeric one, and a small decrease to 60/40 may yield a brittle and plastic product. The optimal ratio must be determined for each case.

Poly(ethylene adipate) may be prepared by the method of Brooks et al. (7). Polyether glycols may be prepared by the techniques of Bailey et al. (8). Polycarbonates were described in detail by Schnell (9) and have been reviewed extensively by Schnell (10) and by Christopher and Fox (11).

Block copolymers not unlike those described here were prepared by Merrill (12,13), Goldberg (14), and Reader and Rulison (15).

Perry, Jackson, and Caldwell (4) used poly(tetramethylene oxide) for most of their experiments but included polyesters, polyformals, and polycarbonates. Elastomers were made with polycarbonate blocks based on bisphenol-A, 4,4'-(2-norbornylidene)bisphenol, its 2,6-dichloro derivative, and 4,4'-(hexahydro-4,7-ethanoindan-5-ylidene)diphenol.

Wismer, Memon, and Williams (6,16) used a commercial poly(ethylene adipate) [Multrathane R14 (Mobay)] and commercial hydroxy-terminated polybutadienes [Butarez HTB (Phillips Petroleum Co.) and R-45M (Sinclair Petrochemicals Inc.)] . In addition to bisphenol-A, they used an equal weight of cyclohexylidine-bisphenol; 2,2-diphenylpropane-1,3-propane-1,3-diol; 2,2,4,4-tetramethyl-1,3-cyclobutanediol; 1,4-cyclohexanedimethanol; 1,4-cyclohexane-diol; or phenolphthalein. An elastomeric polycarbonate was prepared from 1,4-butanediol by the method of Jackson and Caldwell (17).

Chain extension of diols by diisocyanates has been described extensively by Saunders and Frisch (18). The reaction used here is similar to that described by Brooks, Bledsoe, and Rodriquez and checked by Stratton and Ryan (7). It should be recalled that the urethane units will be near the middle of the "terminal" polycarbonate blocks, that is, the urethane units will be surrounded by the glassy polycarbonates which are joined in turn to the elastomeric diol units.

This extension to Perry, Jackson, and Caldwell's procedure was desired so that the block copolymer could be produced with as low a molecular weight as possible for ease of processing but also with as strong and elastic a final product as possible, using a relatively simple and well-tried procedure for chain extension. Loss of thermoplastic properties may result if the diisocyante brings about cross-linking as well as chain extension.

5. References

1. N. A. Memon, Ph.D. Dissertation, 1972, University of Toronto, *Chem. Abstr.* 79, 42877C (1973).
2. H. Leverne Williams, Professor, Department of Chemical Engineering and Applied Chemistry, University of Toronto, Toronto, Ontario M5S 1A4, Canada, to whom all correspondence should be directed.
3. Research Laboratories, Eastman Kodak Co., 1669 Lake Ave., Rochester, New York 14650.
4. K. P. Perry, W. J. Jackson, Jr., and J. R. Caldwell, *J. Appl. Polym. Sci.* 9, 3451 (1965).
5. M. D. Hartley and H. Leverne Williams, *J. Appl. Polym. Sci.* 19, 2431 (1975).
6. N. A. Memon and H. Leverne Williams, *J. Appl. Polym. Sci.* 17, 1361.
7. T. W. Brooks, C. Bledsoe, and F. Rodriquez, checked by G. Statton and T. Ryan, *Macromol. Syntheses*, 4, 1 (1972).
8. F. E. Bailey, Jr., and H. G. France, checked by C. C. Price, R. Spector, and Y. Atarashi, *Macromol. Syntheses* 3, 77 (1968).
9. H. Schnell, *Angew. Chem.* 68, 633 (1956).
10. H. Schnell, *Chemistry and Physics of Polycarbonates*, Wiley, New York, 1964.
11. W. F. Christopher and D. W. Fox, *Polycarbonates*, Reinhold, New York, 1962.
12. S. H. Merrill, *J. Polym. Sci.* 55, 343 (1961).
13. S. H. Merrill and S. E. Petrie, *J. Polym. Sci. Al* 3, 2189 (1965).

14. E. P. Goldberg, *J. Polym. Sci. C* **4**, 707 (1963).
15. A. M. Reader and R. N. Rulison, *J. Polym. Sci. Al* **5**, 927 (1967).
16. K. H. Wismer, B.A. Sc. Dissertation, 1969, University of Toronto. Data quoted in references 1 and 6.
17. W. J. Jackson, Jr., and J. R. Caldwell, *Ind. Eng. Chem. Prod. Res. Dev.* **2**, 246 (1963).
18. J. H. Saunders and K. C. Frisch, *Polyurethanes, Chemistry and Technology, Vol. 1, Chemistry; Vol. 2, Technology*, Interscience, New York, 1965.

Polyimide from β-Carboxymethyl Caprolactam

{Poly[1,4-(2,6-Dioxopiperidinediyl)trimethylene]}

Submitted by H. K. Reimschuessel (1)
Checked by W. A. Bowman (2)

1. Procedure

A. Monomer Synthesis (3)

A mixture of 384 g of α-bromocaprolactam (Note 1) and 400 ml of 2,6-lutidine is heated to reflux with stirring for 3 hr. After cooling, the precipitate is filtered and washed with benzene. (*Caution! Benzene is toxic and should be used in a hood.*) The filtrates are combined. Any additional precipitate which forms is filtered (Note 2).

The filtrate (Note 3) is then added dropwise to a refluxing solution of sodium diethyl malonate which is prepared by reaction of 50.6 g of sodium with 3 liters of absolute ethanol, followed by dropwise addition of 640 g of diethyl malonate and refluxing for 3 hr. After the addition is completed the reaction mixture is refluxed for an additional 5 hr. After cooling, ethanol and benzene are removed by distillation. The residue is dissolved in ether, and the resulting solution is washed with 4% HCl, followed by saturated $NaHCO_3$ solution and finally with water. After drying the ether solution with Na_2SO_4, ether and diethyl malonate are removed by distillation *in vacuo*, and the resulting oily residue is triturated repeatedly with petroleum ether (b.p. 35-50°). The resulting crystalline material is then recrystallized from *n*-hexane. β-(Dicarbethoxymethyl) caprolactam [350 g (65%)] is obtained (m.p. 49-50°).

A solution of 350 g of β-(dicarbethoxymethyl) caprolactam in 500 ml of absolute ethanol is added dropwise to a refluxing solution of 175 g of KOH in 1 liter of absolute ethanol. After addition, refluxing is continued for 6 hr. The precipitate [376 g (dry)] is filtered, washed with ethanol, and dissolved in 250 ml of water. This solution is cooled to about −5° and 241 ml of concentrated HCl is added dropwise. The precipitate is filtered and washed with cool methanol and ether. Yield is 252 g (91%). The material, β-(dicarboxymethyl) caprolactam, melts at 162° with decomposition.

β-(Dicarboxymethyl) caprolactam (20 g) is added in small quantities to *o*-dichlorobenzene which is heated to 160-165°. After CO_2 evolution has ceased, the mixture is heated to 170° (Note 4). The resulting clear solution is then allowed to cool to room temperature; 18.4 g of β-carboxymethyl caprolactam precipitates upon cooling. Recrystallization from water yields a product that melts at 193-194°.

B. Polymerization (7)

1. Ten grams of β-carboxymethyl caprolactam is charged to a polymerization tube fitted with connections for nitrogen and vacuum. The tube is purged of air by alternately evacuating and filling it with nitrogen. The tube is then heated at 200° for 6 hr while a current of nitrogen is passed through the tube. The material is completely converted to a hard, crystalline polymer. A reduced viscosity of 0.6-0.7 dl/g is usually found for a solution of 0.52 g of polymer in 100 ml of *m*-cresol (Notes 5 and 8).

2. The same procedure as under preparation 1 is performed except the poly-

merization temperature is 280°. The polymer is a hard, glossy material having a reduced viscosity in *m*-cresol solution between 2.0 and 3.0. (Notes 6, 7, and 8).

2. Characterization (8)

Solubility: The polymer is soluble in formic acid, *m*-cresol, trifluoroethanol, and sulfuric acid; it is insoluble in common organic solvents (Note 8).

Melting point: The crystalline polymer melts at 279-280°.

Glass transition temperature: 91° by differential thermal analysis.

Spectral analysis: The ir spectrum of a film cast from formic acid solution exhibits strong bands at 1728 and 1677 cm^{-1}.

Degree of polymerization (9): The number average degree of polymerization may be estimated from the reduced viscosities in *m*-cresol (0.52 g of polymer/100 ml), according to $[\eta] = 3.77 \times 10^{-2} (P_n)^{0.68}$, where $[\eta]$ may be obtained from the relation $\eta_{spec}/c = [\eta] + 0.35 [\eta]^2 c$.

3. Notes

1. The α-bromocaprolactam may be obtained as described in references 4-7. Best results are obtained by the methods to which reference 7 refers, which entail the synthesis of 2-chloro-azacyclo-2,3-heptene-*N*-carbochloride followed by addition of bromine.

2. The precipitate, about 370 g, is 2,6-lutidine hydrobromide (m.p. 213°).

3. The filtrate can be used without prior removal of excess lutidine.

4. Any insoluble material may be removed by filtering the hot mixture.

5. The checker obtained a reduced viscosity of 0.52 in a yield of ~90%.

6. The checker obtained a reduced viscosity of 3.2 in a yield of ~90%.

7. Addition of an amine or an anhydride such as succinic anhydride may be added to regulate the viscosity.

8. The checker found the polymer to be difficultly soluble in *m*-cresol.

4. References

1. Allied Chemical Co., P.O. Box 1021R, Morristown, New Jersey 07960.
2. Eastman Kodak Co., 1669 Lake Ave., Rochester, New York 14650.
3. H. K. Reimschuessel, J. P. Sibilia, and J. V. Pasxale, *J. Org. Chem.* **34**, 959 (1969).
4. W. C. Francis, J. R. Thornton, J. C. Werner, and T. R. Hopkins, *J. Am. Chem. Soc.* **80**, 6238 (1958).
5. R. J. Wineman, E. P. Hsu, and C. E. Anagnostopoulos, *J. Am. Chem. Soc.* **80**, 6233 (1958).

6. J. H. Ottenheym and J. W. Garristen, German Patent 1,154,118 *Chem. Abstr.* **60**, 2789e (1959).
7. British Patent 901169, ex. 1, *Chem. Abstr.* **58**, 6810e (1963); U.S. Patent 3000881, ex. 2, *Chem. Abstr.* **56**, 2337d (1962).
8. H. K. Reimschuessel, L. G. Roldan, and J..P. Sibilia, *J. Polym. Sci. A2* **6**, 559 (1968).
9. H. K. Reimschuessel, *Trans. N.Y. Acad. Sci. II* **33**, 2.9 (1971).

Polyimide from 4-Carboxy-2-Piperidone

{Poly[1,3-(2,6-Dioxopiperidinediyl)ethylene]}

Submitted by K. P. Klein and H. K. Reimschuessel (1)
Checked by W. A. Bowman (2)

1. Procedure (3)

A. *Monomer Synthesis*

A 2-liter three-necked round-bottomed flask is equipped with a mechanical stirrer, two addition funnels, and an ice-water cooling bath. In the flask is placed a solution of 79.0 g (0.5 mole) of distilled dimethyl itaconate (Note 1) in 400 ml of methanol. A solution of 65.1 g (1 mole) of potassium cyanide in 400 ml of water is added dropwise from one addition funnel (Note 2). After 0.5 hr, 65 ml (0.80 mole) of cold, concentrated hydrochloric acid is added from the other addition funnel. The reaction mixture is stirred at room temperature for 72 hr and then extracted with three 250-ml portions of ether. The ethereal extracts are combined, dried with $MgSO_4$, and concentrated *in vacuo*. The resulting oil is distilled to afford 51.8-71.4 g (56-77%) of dimethyl cyanomethyl succinate (Note 3).

Dimethyl cyanomethyl succinate [46.3 g (0.25 mole)] is dissolved in 350 ml of ethanol, then 4 g of Raney nickel catalyst (Note 4) is added. The resulting suspension is placed in a 500-ml Parr bomb and hydrogenated at 60 psi and 50° until no further hydrogen uptake is observed. The suspension is removed from the Parr apparatus, filtered, and the filtrate is concentrated *in vacuo*. The resulting solid is recrystallized from methanol by cooling the methanolic solution in a dry-ice-acetone bath. The yield of 4-methoxycarbonyl-2-piperidone is 29.5-34 g (76-86%) (Note 5).

To a stirred solution of potassium hydroxide [12.3 g (0.22 mole)] in 150 ml of methanol in a 500-ml one-necked flask is added 4-methoxycarbonyl-2-piperidone 31.4 g (0.2 mole) dissolved in 150 ml of methanol. The flask is equipped with a reflux condenser and the solution is refluxed for 10 hr. The methanol is then removed *in vacuo*. The resulting solid is dissolved in 50 ml of water, cooled to 0°, and acidified with concentrated hydrochloric acid. Three recrystallizations of the precipitate from ethanol-water yielded 18.5-25.5 g (65-89%) of the monomer, 4-carboxy-2-piperidone (Note 6).

B. Polymerization Procedure

1. 4-Carboxy-2-piperidone (5 g) and 4 drops (0.2 ml) of water is placed in a 20 × 20-mm Pyrex polymerization tube. The tube is swept with nitrogen for 5 min and immersed to a depth of 6 in. in an oil bath maintained at a constant temperature of 220°. After 24 hr the polymerization tube is removed from the bath and allowed to cool at room temperature under a continuous nitrogen sweep. The tube is then broken away from the polymer plug which is ground. After extraction in boiling ethanol with subsequent drying at 100° and 30 torr, the polymer sample exhibits a reduced viscosity value of 0.8 dl/g when measured using 0.13 g in 25 ml of *m*-cresol at 25° (Note 7).

2. Two 5-g samples of 4-carboxy-2-piperidone are heated at 200° and 220°, respectively, for 20 hr as described in example 1 and then heated at constant temperature for an additional 20 hr under vacuum (2 torr). The polymer samples, when treated as described above, give reduced viscosity values of 1.0 and 1.4 dl/g, respectively, when measured using 0.13 g in 25 ml of *m*-cresol at 25° (Note 8). The ir spectra of these samples are identical with those of the polymer from example 1.

3. A 5-g sample of 4-carboxy-2-piperidone is placed in a polymerization tube and the inlet tube is constricted with an oxygen torch. The tube is alternately evacuated with an oil pump, flushed with nitrogen several times, and then sealed under vacuum. The tube is heated at 230-235° for 24 hr. After cooling to ambient temperature, the inlet tube is opened and attached to an oil pump. The tube is then heated at 235° and 0.5 torr for 5 hr. Treatment as in example 1 affords a polymer having reduced viscosity of 0.35-0.60 dl/g.

2. Characterization (4)

Solubility. The polyimide is insoluble in most low-boiling organic solvents, but is soluble in formic acid, *m*-cresol, trifluoroethanol, and sulfuric acid. The checker found the polymers to be very difficult to dissolve in *m*-cresol.

Spectral Analysis. The ir spectrum of a thin polymer film that had been cast from a trifluoroethanol solution exhibited strong absorptions at 1705 and 1790 cm^{-1} assigned to the carbonyl vibrations. A 15% solution of the polyimide in formic acid afforded the following nmr spectrum: δ = 1.17-2.13 (m, NCH$_2$CH), 2.44 (m, COCH$_2$), and 3.26 (m, NCH$_2$) ppm. The chemical shift of the methine proton was obscured by the two broad multiplets at 2.44 and 3.26; the total area of these resonances corresponded to five protons while that at 1.17-2.13 corresponded to two protons.

Viscosity-Molecular-Weight Relations. Solutions of 0.52 g of polymer in 100 ml of either *m*-cresol or trifluoroethanol were used for determining the reduced viscosities. Identical values were obtained for both polymers. Intrinsic viscosities were determined on *m*-cresol solutions and evaluated by using the Huggins equation, $\eta_{spec}/c = [\eta] + k'[\eta]^2c$. A value of 0.41 was determined for k'. The relationship between intrinsic viscosity and the molecular weight was

$$[\eta] = 4.2 \times 10^{-4}\bar{M}_w^{0.647}$$

A glass transition temperature of 130-135° was determined by differential thermal analysis.

3. Notes

1. Dimethyl itaconate was obtained from the Aldrich Chemical Co.

2. *Caution! Adequate precautions should be taken when handling potassium cyanide.* Upon the addition of hydrochloric acid, hydrogen cyanide is produced. This entire preparation step should be carried out under a hood with proper ventilation. Care should be exercised in the extraction step because free hydrogen cyanide is present in the ethereal extracts.

3. Dimethyl cyanomethyl succinate: b.p. 113-117° (0.8 torr); ir (neat) 2260 (C≡N) and 1745 (C=O) cm^{-1}; nmr (CCl$_4$) δ = 2.45-3.40 [m, 5, CH$_2$CH(CO)CH$_2$], 3.68 (s, 3, OCH$_3$), and 3.74 (s, 3, OCH$_2$) ppm. *Analysis.* Calcd. for C$_8$H$_{11}$NO$_4$: C, 51.88%; H, 5.99%; N, 7.57%. Found: C, 51.58%; H, 5.76%; N, 7.76%.

4. The Raney Active Nickel Catalyst (No. 28) was obtained from the W. R. Grace & Co.

5. 4-Methoxycarbonyl-2-piperidone: m.p. 126.5-127°; ir (KBr) 1740 (ester C=O), 1665 (lactam C=O) and 1640 (C=N); nmr (CDCl$_3$) δ = 2.0-3.2 [m, 5, CH$_2$CH(CO)CH$_2$], 3.36 (m, 2, CH$_2$N), 3.73 (s, 3, OCH$_3$), and 7.86 (s, 1, NH) ppm. *Analysis.* Calcd. for C$_7$H$_{11}$NO$_3$: C, 53.49%; H, 7.06%; N, 8.91%. Found: C, 53.33%; H, 7.14%; N, 8.92% The checker obtained a 60% yield of product melting at 123°.

6. 4-Carboxy-2-piperidone: m.p. 174.5-175.5°; ir (KBr) 1695 (acid C=O) and 1635 (lactam C=O) cm^{-1}; nmr (D$_2$O) δ = 1.7-3.4 [m, 5, CH$_2$CH(CO)CH$_2$] and 3.5 (m, 2, CH$_2$N) ppm. *Analysis.* Calcd. for C$_6$H$_9$NO$_3$: C, 50.34%; H, 6.34%; N, 9.79%. Found: C, 50.36%; H, 6.36%; N, 9.60%.
The checker prepared this compound at one-half the reported scale and obtained a yield of 42% melting at 171.5°.

7. The checker obtained a polymer with reduced viscosity of 1.14 dl/g in a yield of about 80% at one-fifth the reported scale.

8. The checker obtained polymers with reduced viscosities of 1.2 and 1.40 dl/g, respectively, in a yield of about 80% at one-fifth the reported scale.

4. References

1. Allied Chemical, P.O. Box 1021 R, Morristown, New Jersey 07960.
2. Eastman Kodak Co., 1669 Lake Ave., Rochester, New York 14650.
3. H. K. Reimschuessel, K. P. Klein, and G. J. Schmitt, *Macromolecules* 2, 567 (1969).
4. H. K. Reimschuessel and K. P. Klein, *J. Poly. Sci. A1* 9, 3071 (1971).

Alternating Benzofuran-Maleic Anyhdride Copolymer

{Poly[2,3-(2,3-Dihydrobenzo[b]furandiyl)-3,4-(2,5-Dioxo-oxolanediyl)]}

Submitted by A. Priola and S. Cesca (1)
Checked by J. Charles (2)

1. Procedure

A 50-ml two-necked flask equipped with a nitrogen inlet and outlet is flushed with dry nitrogen for 15 min and then is charged with a mixture of 6.0 g (61 mmole) of maleic anhydride (Note 1), 10 ml of chlorobenzene (Note 2), 7.0 ml (64 mmole) of benzofuran (Note 3), and 0.058 mmole of AIBN (azobisisobutyronitrile) (Note 4).

The reaction vessel is closed with two screw clamps (Note 5) and is then connected to a shaker (Note 6). The reaction flask is placed in an oil bath maintained at 55 ± 1° and is shaken for 20 hr. During this time the copolymer gradually separates. The final product is a pale yellow, compact mass swollen with solvent.

The contents of the flask are poured into 200 ml of toluene and the precipitate is filtered through a fine-fritted filter under dry nitrogen. The solid mass of polymer is washed twice with toluene, then with dry diethyl ether, and finally is recovered and dried under reduced pressure (0.2 torr) at room temperature for 15 hr.

The copolymer is a white powder and weighs 7.4 g (yield 54.5%, Note 7). The crude copolymer can be further purified by dissolution in methyl ethyl ketone (MEK) and precipitation into toluene (volume ratio of toluene:polymer solution = 5:1).

2. Characterization

The purified copolymer is soluble in polar solvents such as acetone, MEK, cyclohexanone, dioxane, tetrahydrofuran, *N,N*-dimethylformamide, and *N,N*-dimethylacetamide. It is insoluble in H_2O at pH = 7, but after hydrolysis of the anhydride units it is water-soluble at pH $\geqslant$ 11.

The copolymer has an intrinsic viscosity $[\eta]$ = 1.0 dl/g in MEK at 30°. The number average molecular weight is $\bar{M}_n$ = 214,000 (Note 8).

Differential thermal analysis (Note 9) of the copolymer does not show the presence of any phase transition in the temperature range 0-290°. The copolymer decomposition starts at 290°.

The elemental analysis of the copolymer approaches the calculated values for equimolar composition: Found: C, 65.8%; H, 4.1%. Calcd. C, 66.7%; H, 3.7%.

Infrared spectra (KBr disk) (Note 10) show the presence of bands at 1730, 1780, and 1860 cm^{-1} from maleic anhydride units, and bands at 752, 1235, 1462, 1480, 1595, and 1610 cm^{-1} from benzofuran units [determined by comparison with the ir spectrum of benzofuran homopolymer (3)].

H-nmr analysis (Note 10) gave poorly resolved spectra. However, a signal at δ = 3.6 ppm, from the protons of structural units derived from maleic anhydride, and a signal at δ = 6.8 ppm, attributable to the aromatic protons of benzofuran, are evident. The quantitative evaluation of the areas under these signals agreed with the equimolar composition of the copolymer.

The alternating structure of the copolymer is characterized by ^{13}C-nmr (Note 10) (Fig. 1) and, indirectly, is supported by the inability of the individual monomers to homopolymerize under the experimental conditions described above. The characteristic features of *Fig. 1* are as follows (5):

1. The aliphatic carbon atoms resonate in the region from δ = 45 to 100 ppm. The signals from C_3, C_{10}, and C_{11} are centered around 48 ppm, while the proximity of the oxygen atom moved the sharp signal of C_2 downfield (85 ppm).

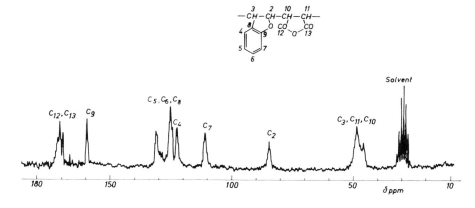

Fig. 1. Fourier transform ^{13}C-nmr spectrum of alternating maleic anhydride-benzofuran copolymer in $(CD_3)_2CO$ at room temperature. Chemical shift of δ in ppm from TMS.

2. The signals of the aromatic carbon atoms are evident in the range from δ = 100 to 160 ppm. They are rather sharp and appear at 112 (C_7) and 123 (C_4), and from 125 to 131 (C_5, C_6, and C_8) ppm, whereas the C_9 signal is shifted to 159 ppm, analogously to the C_2 signal.

3. The carbonyl resonance appears as three peaks at δ = 169.5, 171, and 172 ppm.

3. Notes

1. Maleic anhydride was purified by crystallization from $CHCl_3$ and subsequent sublimation under vacuum (0.2 torr) at 52-54°.

2. Chlorobenzene is preferably freshly distilled under vacuum (10 torr) from CaH_2. Solvents such as cyclohexanone (4) or dioxane and toluene (5) can be used instead of chlorobenzene, but in these cases the copolymer shows lower molecular weight.

3. Benzofuran was purified by distillation under vacuum from CaH_2. Gas chromatographic analysis (diethyl succinate column, 5% on chromosorb; length = 5 m; T = 120°; carrier = He 1 ml/10 sec) indicated the presence of one unidentified impurity (0.4% by area) which had a retention time lower than that of benzofuran.

4. AIBN is purified by crystallization from ethanol, particularly when it is not a fresh product.

5. SOVIREL Catalog 1973, p. 101. The reaction can also be carried out in a beverage bottle stoppered with a crown cap, or in a large vial, sealed with a flame

under vacuum after degassing the reagents by several freeze-thaw cycles at dry ice or liquid N_2 temperatures.

6. A conventional mechanical stirrer provides insufficient agitation of the reaction mixture as the reaction proceeds and the polymer separates. An amount of solvent greater than that described here would allow easier agitation, but the resulting copolymer shows a lower molecular weight.

7. The checkers obtained a yield of 31%.

8. The number average molecular weight was determined with a Mechrolab instrument (model 502) with anhydrous MEK (distilled under a nitrogen atmosphere) as solvent.

9. A DuPont instrument (model 900) equipped with a DSC cell ($16°$/min) was used.

10. Infrared analysis was carried out with a Perkin-Elmer (model 125) spectrometer. A Varian H-100 spectrometer was used for H-nmr analysis (TMS as internal standard; solvent acetone-d_6). ^{13}C-nmr spectra were recorded at 22.63 MHz in the pulsed Fourier transform mode with a HFX-Bruker spectrometer equipped with a 1083 Nicolet computer.

4. Methods of Preparation

Equimolar copolymers of benzofuran and maleic anhydride have been prepared in cyclohexanone using AIBN as initiator (4). Under these conditions the highest reduced viscosity was $\eta_{red} = 0.11$ dl/g in acetone at $25°$.

Our investigations (5) have shown that high monomer concentration does not increase the molecular weight of copolymer beyond the value of $[\eta] = 0.20$ dl/g when cyclohexanone is used as solvent. The same results were obtained when the solvent was dioxane or toluene instead of chlorobenzene (5).

5. References

1. Snamprogetti S.p.A., Polymer Research Laboratories, San Donato Milanese 20097, Italy.
2. GAF Corporation, 1361 Alps Road, Wayne, New Jersey 07470.
3. G. Natta, M. Farina, M. Peraldo, and G. Bressan, *Makromol. Chem.* 43, 68 (1961).
4. E. J. Goethals, A. Cardon, and R. Grosjean, *J. Macromol. Sci. Chem.* A7(6), 1265 (1973).
5. S. Cesca, G. Gatti, and A. Priola, *Chim. Ind. (Milan)* 58, 317 (1976).

Polyterephthaloyl-oxalamidrazone

{Poly[Iminonitrilo(diaminoethanediylidene)-nitriloiminoterephthaloyl]}

Submitted by M. Wallrabenstein (1)

Checked by W. L. Hofferbert, Jr., and J. Preston (2)

Polyterephthaloyloxalamidrazone is prepared in high yields by the reaction of oxalamidrazone with terephthaloylchloride by the interfacial method (3).

1. Preparation

In a 250-ml Erlenmeyer flask 1.39 g of oxalamidrazone (Note 1) is dissolved in 200 ml of oxygen-free water (Note 2) at 25-30° under nitrogen purge and magnetic stirring. The stirring bar is removed and the solution is transferred to a blender. Subsequently a solution of 2.02 g of sodium bicarbonate, dissolved in 50 ml of oxygen-free water, is added. The agitator vessel is cooled with running water. Under vigorous stirring (13,000 rpm) a solution of 2.44 g of terephthaloylchloride (Note 3) in 250 ml of dried hexane (Note 4) is immediately poured into the blender. Stirring is continued for 6 min and the polymer suspension is suction-filtered. The extensively swollen filter cake is made into a paste with 200 ml of water with manual stirring and filtered. This washing process is repeated twice and methanol is used as the final wash. The polymer is dried overnight in a vacuum drier at 20 torr and 70°. A yellow polymer in a yield of 94-96% is obtained.

2. Characterization

Polyterephthaloyloxalamidrazone is very slightly soluble in organic solvents. It is readily soluble in alkaline media (4), for example, sodium and potassium hydroxide solutions (with deep red coloration), and in sulfuric acid and methanesulfonic acid.

The inherent viscosity of the polymer is η_{inh} = 1.10-1.15 dl/g (0.5 g/dl of a 3% KOH solution at 20°).

The ir spectrum of polyterephthaloyloxalamidrazone shows characteristic absorption bands at 3410 (m), 3320 (m), 3190 (m) cm^{-1} for NH; 1630 (s) and 1530 (s) cm^{-1}, 1600 (s) and 1495 (m) cm^{-1}, 1320 (m) and 1275 (m) cm^{-1} for CONH.

3. Notes

1. Oxalamidrazone is prepared by the reaction of cyanogen with hydrazine hydrate (5) in ethanol at -10°. The product is washed repeatedly with ethanol and recrystallized from pure dimethylformamide under nitrogen. The product is rewashed with anhydrous methanol and dried at 3 torr and 40°.

2. Demineralized water is always used. It is deoxygenated by boiling for 1 hr while nitrogen is passed through and then cooling to room temperature.

3. The submitters used terephthaloylchloride from Dynamit Nobel, Troisdorf (Germany). It is purified by two fractional distillations through a 40-cm Vigreux column under vacuum at 15 torr and 138°.

4. Hexane is dried and stored over Na wire.

4. Other Methods of Preparation

Further variants of the interfacial polycondensations which increase the viscosity values of the polymer considerably are:

1. The use of sodium or lithium chloride (6) (10-15% relative to the quantity of water used).

2. Polycondensation at lower monomer concentrations.

5. References

1. AKZO Research & Engineering, Research Laboratories, 8753 Obernburg, Germany.
2. Monsanto Triangle Park Development Center, Inc., Research Triangle Park, North Carolina 27709.
3. M. Saga and T. Shono, *Polymer Lett.* **4**, 869 (1966); Enka Glanzstoff AG, DT-OS 1595775 correspond. U.S. Patent 3544528, *Chem. Abstr.* **74**, 55008s (1971).
4. Enka Glanzstoff AG, DT-Patent 1694324 correspond. U.S. Patent 3583953, *Chem. Abstr.* **71**, 125463a (1969).
5. T. Curtius and G. M. Dedichen, *J. prakt. Chem.* **50**, 245, 253 (1894); *P. M. Hergenrother Polymer Preprints* **13**(2), 930 (1972).
6. W. Wallrabenstein, A. Schoepf, and D. Frank, Enka Glanzstoff AG, DT-OS 1950907 correspond. U.S. Patent 3876586, *Chem. Abstr.* **75**, 21499h (1971); Enka Glanzstoff AG, DT-OS 1950908 correspond. U.S. Patent 3718625, *Chem. Abstr.* **75**, 21477z (1971).

Sulfone-Containing Poly (Amide-Imides)

{Poly[2,5-(1,3-Dioxoisoindolinediyl)sulfonyl-1,4-Phenylenecarbonylimino-4,4'-Biphenyldiyl]}

Submitted by J. M. Adduci and S. K. Sikka (1)
Checked by R. A. Brand and J. E. Mulvaney (2)

1. Procedure

A. *3,4,4'-Tricarboxydiphenyl Sulfone (3)*

To a suspension of 20 g (0.077 mole) of 3,4,4'-trimethyldiphenyl sulfone (Note 1) in 600 ml of water, 80 ml of pyridine is added. The mixture is refluxed and 162 g (1.12 mole) of potassium permanganate is added in five portions (*Caution!* Note 2). After all the permanganate is added, the heterogeneous mixture is refluxed for 1 hr. The mixture is cooled, the manganese dioxide is removed by filtration, and the filtrate is concentrated to about 400 ml. The tricarboxydiphenyl sulfone is precipitated with hydrochloric acid, filtered, washed with water, and dried to give 24.5 g (92% yield). Recrystallization from acetonitrile gives m.p. 270-271°.

B. *3,4-Dicarboxy-4'-(Chloroformyl)diphenylsulfone Anhydride (3)*

A suspension of 3.5 g (0.10 mole) of 3,4,4'-tricarboxydiphenyl sulfone in 50 ml of benzene and 2.38 g (0.20 mole) of thionyl chloride is refluxed for 20 hr. The benzene and unreacted thionyl chloride are removed under vacuum (aspirator), leaving a residue of 3.4 g (97% yield) of 3,4-dicarboxy-4'-(chloroformyl)diphenylsulfone anhydride (Note 3). Recrystallization from benzene gives m.p. 200-201°.

C. *Polymerization*

A 100-ml round-bottomed three-necked flask (Note 4) fitted with a mechanical stirrer, water-cooled condenser with drying tube, and a nitrogen gas inlet is charged with 1.38 g (0.0075 mole) of benzidine (*Caution!* Note 5) and 23 ml of N,N-dimethylacetamide (Note 6). The solution is cooled to below -15° using an ice-salt-water bath, then 2.630 g (0.0075 mole) of 3,4-dicarboxy-4'-(chloroformyl)diphenylsulfone anhydride is added in one portion. The mixture is stirred for 1 hr at this temperature and then allowed to reach 0°. Pyridine (0.6 g) is added to neutralize the hydrogen chloride formed in the reaction. The mixture is stirred for an additional 6-12 hr and allowed to reach room temperature. This procedure yields a dark yellow, viscous solution which can be used for preparing polyamic acid samples and/or poly(amide-imide) films.

2. Characterization

A polyamic acid sample is used for molecular weight determination and is prepared by triturating the reaction mixture in dry acetone and drying *in vacuo*.

The polyamic acid prepared by this procedure has a number average molecular weight, $\bar{M}_n$ = 53,500 (Note 7).

Poly(amide-imide) films are prepared by casting the polyamic acid solution onto a glass plate using a 15-mil spacer and heating in an oven, gradually increasing the temperature to remove the solvent, and heating *in vacuo* for 3 hr at 200° to insure complete imidization. A transparent, dark yellow, flexible film is obtained by this procedure. The ir spectrum shows absorptions at 3300-3400 (broad), 1650-1680, and 1790 cm^{-1}, consistent with the poly(amide-imide) structure.

Thermal gravimetric analysis of the film (Note 8) shows small weight loss (1-3%) at low temperatures. Polymer weight loss is gradual until 480° and then loss is rapid (9% at 485°, 31% at 579°, and 100% at 673°).

An isothermal weight loss of 38% is observed after 17 days at 315°. After this treatment the poly(amide-imide) film sample could be folded several times before breaking.

The poly(amide-imide) product is insoluble in water, acetone, and *N,N*-dimethylacetamide.

3. Notes

1. The starting material can be prepared by a Friedel-Crafts reaction of *p*-toluene sulfonyl chloride and *o*-xylene using the method of Burton and Praill (4) or Holt and Pagdin (5).

2. Excessive addition of potassium permanganate causes an exotherm and a vigorous reaction. The checkers found it necessary to add the potassium permanganate in many 5-10 g portions to avoid a vigorous reaction.

3. The checkers obtained better yields (55% pure) using thionyl chloride, rather than thionyl chloride-benzene, for the reaction.

4. The flask is flamed and flushed with dry nitrogen. The polymerization reaction is carried out under a dry nitrogen atmosphere.

5. Benzidine is recrystallized from ethanol and melted at 125-126°. Other aromatic diamines may also be used (6).

6. *N,N*-dimethyl acetamide is distilled from calcium hydride and stored over molecular sieves before use.

7. A Hewlett-Packard 502 high-speed membrane osmometer is used to determine molecular weights. The checkers found an inherent viscosity (c = 0.50 g/ml, DMAC) of 0.68 dl/g.

8. The film sample is analyzed in air using a DuPont 950 thermogravimetric analyzer and a heating rate of 20°/min.

4. Methods of Preparation

The low-temperature method used in this procedure has been reported (7). The imide ring closure reaction can also be achieved by chemical methods (8). Reactions with tricarboxylic acids or their derivatives with diisocyanates at 80-100° also give the poly(amide-imide) structure (9,10).

5. References

1. Department of Chemistry, Rochester Institute of Technology, Rochester, New York 14623.
2. Department of Chemistry, University of Arizona, Tucson, Arizona 85721.
3. J. M. Adduci, R. S. Ramirez, F. R. Diaz, and F. Horn, *Rev. Latinoamer. Quim.* **2**, 121 (1972).
4. H. Burton and P. F. G. Praill, *J. Chem. Soc.*, 887 (1955).
5. G. Holt and B. Pagdin, *J. Chem. Soc.*, 2508 (1960).
6. J. M. Adduci, S. K. Sikka, L. E. Migueles, and R. S. Ramirez, *J. Polym. Sci.* **11**, 1321 (1973).
7. W. Wrasidlo and J. M. Augl, *J. Polym. Sci. A1*, **7**, 321 (1969).
8. A. K. Boxe, F. Greer, and C. C. Price, *J. Org Chem.* **23**, 1335 (1958).
9. J. Sambeth, French Patent 1,498,015 (Oct. 13, 1967); *Chem. Abstr.* **69**, 67926z (1968).
10. E. G. Redman and J. S. Skinner, French Patent 1,501,198 (Nov. 10, 1967); *Chem. Abstr.* **69**, 87799p (1968).

Polystyrene Graft Copolymers on Polybutadiene

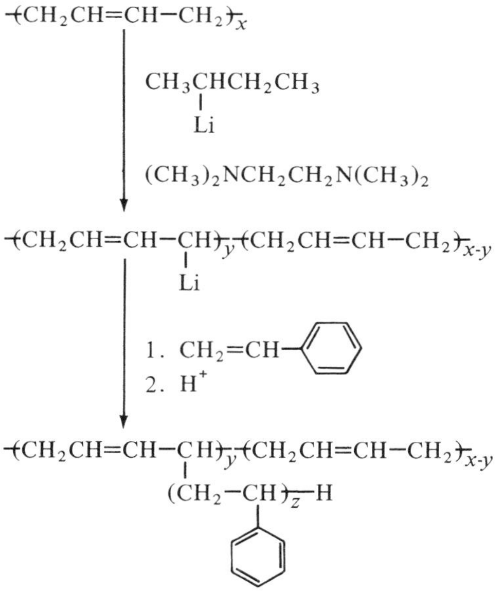

$$+CH_2CH=CH-CH_2\!+_x$$

$$CH_3CHCH_2CH_3$$
$$|$$
$$Li$$

$$(CH_3)_2NCH_2CH_2N(CH_3)_2$$

$$+CH_2CH=CH-CH+_y+CH_2CH=CH-CH_2+_{x-y}$$
$$|$$
$$Li$$

1. $CH_2=CH-\!\!\bigcirc$
2. H^+

$$+CH_2CH=CH-CH+_y+CH_2CH=CH-CH_2+_{x-y}$$
$$|$$
$$(CH_2-CH+_z-H$$
$$\bigcirc$$

Submitted by J. C. Falk, J. Van Fleet,
D. F. Hoeg, J. F. Pendleton, and R. J. Schlott (1)
Checked by G. A. Moczygemba and H. L. Hsieh (2)

1. Procedure (Note 1)

A. Determination of sec-Butyllithium Concentration

Cyclohexane (25 ml) is added to a serum-stoppered 50-ml Erlenmeyer flask. Indicator solution is added [1.0 ml (1.0 g/100 ml of o-phenanthroline in

57

benzene)], followed by 1.0 ml of *sec*-butyllithium (Note 2). The solution turns bright red. The solution is titrated to yellow with a 1.0 *M* solution of isopropanol in cyclohexane. This conditioning is done twice prior to the actual determination and is required to give a sharp end point. The *sec*-butyllithium solution of unknown concentration is added (1.0 ml), and the red color is titrated to the yellow end point with 1.0 *M* isopropanol in cyclohexane (1.5 ml is required). This is repeated three times, [*sec*-butyllithium] = 1.5 *M*.

B. *Polystyrene Graft Copolymers on Polybutadiene (Note 3)*

Polybutadiene (120,000 molecular weight, 25.0 g) in 800 ml of cyclohexane is placed in a capped quart beverage bottle which has been dried overnight at ~120° and cooled under nitrogen. The system is purged through a long hypodermic needle with Matheson pre-purified nitrogen for 10 min. Tetramethylethylenediamine (2.08 mmole) and *sec*-butyllithium [2.08 mmole (Note 4)] are added with a hypodermic syringe (Note 5). The reaction mixture is allowed to stand at room temperature for 2 hr. Styrene [18.1 g (0.174 mole, 20.0 ml)] is added with a hypodermic syringe to the reaction mixture. The reaction mixture is heated at 50° for 6 hr. The amber color of living polystyrene is titrated to colorless with 1.0 *M* isopropanol in cyclohexane (theoretical 2.08 ml; 1.9 ml is actually required). The polymer solution is coagulated in methanol (one volume of polymer solution to four volumes of methanol) and dried at 50° in a vacuum oven (~3 torr) overnight to yield 43 g (100%).

2. Characterization

The single-point solution viscosity determined in toluene at 25° with a Cannon-Ubbelohde viscometer is 1.46 dl/g (0.1 g/dl).

Extraction of the graft copolymer (1.5 g) in a Soxhlet apparatus with 200 ml of acetone, a polystyrene solvent, for 1 hr gives graft efficiency data. The percent graft efficiency is determined from the following equation:

$$\text{Percent graft Efficiency} = \frac{(\text{Total weight of styrene polymerized}) - (\text{Weight of polystyrene extracted with acetone})}{(\text{Total weight of styrene polymerized})} \times 100$$

Graft efficiency = 96%

A film of the rubber is cast by pouring a 30% (w/w) solution of the rubber in methylene chloride into an aluminum foil dish. The solvent is evaporated at room temperature and atmospheric pressure for two days. Final drying is done at 50° under vacuum (~3 torr) for 10 hr. The rubbery film is peeled from the dish.

3. Notes

1. Styrene (Monsanto) is distilled under vacuum (~7 torr) and used immediately. Cyclohexane (McKesson) is passed over molecular sieves prior to use. Polybutadiene is prepared by anionic polymerization in cyclohexane and has a GPC polydispersity index of 1.1. *Sec*-butyllithium may be purchased from the Foote Chemical Co.

2. *Sec*-butyllithium must be stored below 0° or it rapidly loses its activity.

3. All reagents and reactions are kept under a nitrogen atmosphere.

4. This quantity of *sec*-butyllithium corresponds to an average of ten metalated sites per polybutadiene chain.

5. The reaction mixture will, at this point, contain some impurities which will react with *sec*-butyllithium. Therefore, additional *sec*-butyllithium must be added to purge the system. This may be done in two ways: (a) *sec*-Butyllithium is added dropwise until a yellow color occurs. Then the requisite quantity for metalation is added. (b) If the polybutadiene solution is faintly yellow, add to a second reaction mixture an excess of *sec*-butyllithium and 2 ml of styrene. An amber color develops after a few minutes. Titrate to colorless or to faint yellow with a known amount of isopropanol. The difference between the *sec*-butyllithium added and the isopropanol titer is then added to the actual reaction mixture.

6. If the polybutadiene contains appreciable amounts of lower-molecular-weight polymer, the grafting efficiency determined in this way will be lower because the lower-molecular-weight fraction grafted with high-molecular-weight polystyrene is soluble in acetone.

4. References

1. R. C. Ingersoll Research Center, Borg-Warner Corporation, Wolf and Algonquin Roads, Des Plaines, Illinois 60018.
2. Phillips Petroleum Co., Bartlesville, Oklahoma 74004.

Polystyrene-Polyisoprene-Polystyrene

{α-Hydro-ω-1-sec-Butyl[Poly(1-Phenylethylene)—Poly-(2-Methyl-2-Butenylene)—Poly(1-Phenylethylene)]}

Submitted by J. C. Falk, M. A. Benedetto,
J. Van Fleet, and L. Ciaglia (1)

Checked by G. A. Moczygemba and H. L. Hsieh (2)

1. Procedure

A. Materials

Styrene (Monsanto) is distilled under vacuum (∼7 torr) and used imme-diately. Cyclohexane (McKesson) is passed over molecular sieves prior to use. Isoprene (Eastman Kodak) is distilled at atmospheric pressure through a 2-ft-long column packed with glass helices prior to use. *Sec*-butyllithium is purchased from the Foote Chemical Co. (Note 1).

B. Determination of sec-Butyllithium Concentration

Cyclohexane (25 ml) is added to a serum-stoppered 50-ml Erlenmeyer flask. Indicator solution is added [1.0 ml (1.0 g/100 ml of *o*-phenanthroline in benzene) (*Caution! Benzene is toxic and should be used in a hood.*)], followed by 1.0 ml of *sec*-butyllithium. The solution turns bright red. The solution is titrated to yellow with a 1.0 *M* solution of isopropanol in cyclohexane. This conditioning is done twice prior to the actual determination and is required to give a sharp end point. The *sec*-butyllithium solution of unknown concentration is added (1.0 ml) and the red color is titrated to the yellow end point with 1.0 *M* isopropanol in cyclohexane. This titration is repeated three times, [*sec*-butyl-lithium] = 1.5 *M*.

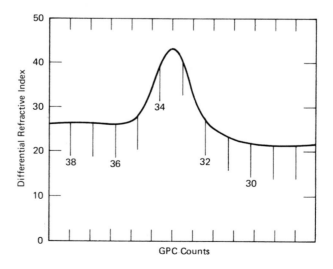

Fig. 1.   Gel permeation chromatogram.

C. Preparation of Polystyrene-Polyisoprene-Polystyrene (Note 2) of 60,000 Molecular Weight

Cyclohexane (600 ml) is added with a hypodermic syringe to a capped quart beverage bottle which had been dried overnight at ~120° and cooled under nitrogen. The system is purged through a long hypodermic needle with Matheson pre-purified nitrogen for 10 min. Styrene (20 g) is added with a hypodermic syringe until a faint yellow color is observed. Immediately, the requisite amount of 1.0 *M sec*-butyllithium (1.66 ml) is added. The reaction mixture is placed in a 70° water bath for 1 hr. Freshly distilled isoprene (60.0 g) is added with a hypodermic syringe. The reaction mixture is kept at 70° for 1 hr. Styrene (20 g) is added and the reacting mixture is heated at 70° for an additional hour. The orange color characteristic of "living" polystyrene is titrated to colorless with 1.0 *M* isopropanol in cyclohexane. The theoretical amount needed for a molecular weight of 60,000 is 1.66 ml. The polymer solution coagulated in methanol (one volume of polymer solution to four volumes of methanol) and dried at 50° in a vacuum oven (~3 torr) overnight to yield 98 g (98%).

2. Characterization

The single-point solution viscosity, determined in toluene at 25° with a Cannon-Ubbelohde viscometer is 0.50 (0.1 g/dl).

The result of gel permeation chromatography is shown in Figure 1. The symmetrical peak indicates that few "living" anions are destroyed during the sequential polymerization.

A film of the rubber is cast by pouring a 30% (w/w) solution of the rubber in methylene chloride into an aluminum foil dish. The solvent is evaporated at room temperature and atmospheric pressure for two days. Final drying is done at 50° under vacuum (~3 torr) for 10 hr. The rubbery film is peeled from the dish.

3. Notes

1. *Sec*-butyllithium must be stored below 0° or it rapidly loses its activity.
2. All reagents and reactions were kept under a nitrogen atmosphere.

4. References

1. Roy C. Ingersoll Research Center, Wolf and Algonquin Roads, Des Plaines, Illinois 60018.
2. Phillips Petroleum Co., Bartlesville, Oklahoma 74004.

Isotactic Copolymer of Acetaldehyde and n-Butyraldehyde

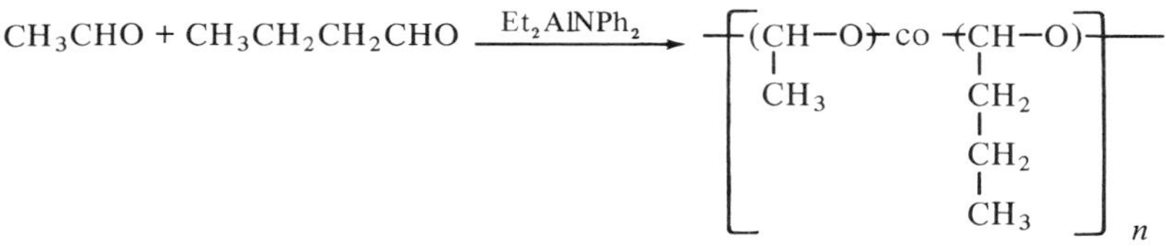

$$CH_3CHO + CH_3CH_2CH_2CHO \xrightarrow{Et_2AlNPh_2} \left[\begin{array}{c} -(CH-O)-co-(CH-O)- \\ \quad | \qquad\qquad\quad | \\ \quad CH_3 \qquad\qquad CH_2 \\ \qquad\qquad\qquad\quad | \\ \qquad\qquad\qquad CH_2 \\ \qquad\qquad\qquad\quad | \\ \qquad\qquad\qquad CH_3 \end{array}\right]_n$$

Submitted by K. Hatada (1)

Checked by R. N. MacDonald (2)

1. Procedure

Caution! Triethylaluminum is highly incendiary and spontaneously ignites in air.

A. The Catalyst

A 200-ml four-necked flask is fitted with a mechanical stirrer, a reflux condenser, a 100-ml addition funnel, and a three-way stopcock. The glassware is dried in an oven at a temperature higher than 100°, assembled while warm, and immediately purged with dry nitrogen. The flask is then charged with 33 ml of dry toluene (Note 1) and 14.5 g of triethylaluminum (17.3 ml, 127 mmole) with hypodermic syringes (Note 2) through the stopcock. Under a nitrogen atmosphere a solution of 21.4 g (127 mmole) of diphenylamine (Note 3) in 40 ml of toluene is added dropwise from the addition funnel over a period of 30 min with agitation, while cooling the flask with an ice-water bath. A fair amount of ethane

65

is released from the reaction of triethylaluminum and diphenylamine. After the addition is complete, the ice-water bath is removed and stirring is continued for 2 hr at 70°. The mixture is then allowed to cool to room temperature. The solution of diethylaluminum diphenylamide thus obtained is diluted with toluene to 0.5 mole/liter on the basis of the triethylaluminum charged. It is stored in an ampoule with a three-way stopcock under a nitrogen atmosphere.

B. *Polymerization*

Procedure A.   A 50-ml glass ampoule is fitted with a three-way stopcock and is subjected to repeated evacuation under flaming with vacuum release by dry nitrogen (Note 4). To the reaction vessel 2.64 g (60 mmole) of acetaldehyde (Note 5) in 20 ml of toluene and 2.88 g (40 mmole) of *n*-butyraldehyde (Note 6) are added with hypodermic syringes through the stopcock. The mixture is cooled to $-78°$ in a dry-ice-acetone bath, and 0.6 ml of catalyst solution is injected slowly with stirring (Note 7). The stopcock is then closed and is allowed to stand at $-78°$. All operations should be carried out under nitrogen. After 5 hr the ampoule is opened (Note 8) and the reaction mass is quickly transferred to a 1-liter blender containing 300 ml of anhydrous ammoniacal methanol (about 4%) precooled at about $-70°$. The blender is turned to high speed with a rheostat for about 20 sec (Note 9) and the mixture is allowed to stand overnight at room temperature. The polymer is collected by filtering through a sintered glass funnel, washed five times with 50-ml portions of methanol, and dried *in vacuo* at 30° for 10 hr. The yield is 2.6-3.0 g.

The content of acetaldehyde in this copolymer is shown to be 62-65 mol. % by an elemental analysis (Note 10).

The copolymer is soluble in some organic solvents such as benzene, toluene, xylene, ethylbenzene, cyclohexane, chloroform, and carbon tetrachloride, although the solubility is fairly low and the resultant solution is highly viscous (Note 11).

Isotactic copolymers of acetaldehyde and *n*-butyraldehyde prepared by this procedure show a typical crystalline powder diffraction pattern with strong scattering from spacings at 4.15 and 8.42 Å (Note 12).

Generally the polymers of aldehydes are thermally unstable (Note 13). A small amount of the copolymer is placed in a test tube and is heated with a small flame of a gas burner while evacuating through a trap cooled in a dry-ice-acetone bath. The copolymer is gently decomposed and the resulting aldehydes can be completely recovered from the trap.

Procedure B.   An alternative procedure can be used to prepare the copolymer.

Here, the catalyst is added to the solution of aldehydes precooled at $-50°$ and the mixture is allowed to stand at $-50°$ for 10 min (Note 14). The temperature of the reaction mixture is then lowered to $-78°$. Polymerization occurs immediately and the solution soon solidifies. After the mixture has stood for 1 hr at $-78°$ the polymerization is stopped. The polymer is isolated and purified as indicated above. The yield is 3.5-4.0 g. The content of acetaldehyde in the copolymer is 63-66 mol. %.

The reaction described here is applicable to the copolymerization of other aldehydes (3). High inherent viscosities may be obtained and the actual values are subject to catalyst concentrations and impurities that may be present.

2. Notes

1. Toluene is purified by fractionating it through a 20-cm Vigreux column, refluxed over calcium hydride for 2 hr, and then distilled under vacuum.

2. The syringe should be greased with a minimum amount of liquid paraffin to prevent the entrance of air. The paraffin is dried over sodium wire for a week.

3. Diphenylamine is purified by recrystallization from *n*-heptane and dried in a desiccator over calcium hydride.

4. Nitrogen gas is dried by passage through a trap packed with 4A molecular sieves cooled in a dry-ice-acetone bath.

5. Acetaldehyde is prepared by depolymerizing pure, dry paraldehyde with anhydrous cupric sulfate as a catalyst and is distilled through a 30-cm Vigreux column. The fraction boiling at $20.5-21.0°$ is collected in an ice-cooled receiver. Traces of water and acid are removed by refluxing over calcium hydride for 2 hr under nitrogen, followed by vacuum distillation. The purification should be carried out just before use.

6. Commercial *n*-butyraldehyde is purified as was acetaldehyde.

7. The reaction mixture becomes orange-red in color, which gradually disappears during the polymerization.

8. The operation should be carefully done wearing leather gloves to prevent injury.

9. Vigorous stirring is needed to isolate the polymer as fine flakes free from catalyst residues, but prolonged stirring sometimes causes the depolymerization of the polymer resulting in reduction of molecular weight.

10. Assuming that the contribution of the depropagation reaction is negligible in this copolymerization, the monomer reactivity ratios were determined to be: $r_1 = 1.97$, $r_2 = 0.59$ at $-78°$ and $r_1 = 1.33$, $r_2 = 0.88$ at $-60°$ (M_1 = acetaldehyde) (4).

11. Many crystalline homopolymers of aldehydes are insensitive to most common organic solvents, but the crystalline isotactic copolymer of acetaldehyde and *n*-butyraldehyde containing 50-80 mol. % of acetaldehyde is soluble in some organic solvents (3,4).

12. The isotactic copolymers of acetaldehyde and *n*-butyraldehyde prepared with diethylaluminum diphenylamide are cocrystalline over the entire composition range, although the crystallinities are slightly lower than those of both homopolymers. The unit cells of these crystalline copolymers have the same tetragonal $I4_1/a$ space group with the same identity period ($C=4.8$ Å) as each homopolymer, while the lattice constant, a, changes continuously with the copolymer composition. From these results it is concluded that isomorphism of monomeric units (5,6) occurs in the case of these copolymers (3,7). "Isomorphism of monomeric units" means that the different monomeric units are incorporated statistically into the chain of the copolymer without disrupting crystallinity itself, causing only continuous variations in the lattice dimensions.

13. The thermal stability of aldehyde polymers is improved by copolymerization, and the copolymer of acetaldehyde and *n*-butyraldehyde prepared here is more stable than either homopolymer (3,4). The stability of the copolymer is further improved if catalyst residues are removed by refluxing it with methanol containing acetylacetone (4,8).

3. References

1. Department of Chemistry, Faculty of Engineering Science, Osaka University, Toyonaka, Osaka, Japan.
2. Central Research and Engineering Department, E. I. DuPont Co., Experimental Station, Wilmington, Delaware 19898.
3. A. Tanaka, Y. Hozumi, K. Hatada, and R. Fujishige, *Kobunshi Kagaku* 20, 694 (1963); *Chem. Abstr.* 60, 14615e (1964).
4. A. Tanaka, Y. Hozumi, and K. Hatada, *Kobunshi Kagaku* 22, 216 (1965); *Chem. Abstr.* 63, 14993f (1965).
5. G. Natta, *Makromol. Chem.* 35, 94 (1960).
6. G. Natta, P. Corradini, D. Sianesi, and D. Morero, *J. Poly. Sci.* 51, 527 (1961).
7. A. Tanaka, Y. Hozumi, K. Hatada, S. Endo, and R. Fujishige, *Polymer Lett.* 2, 181 (1964).
8. A. Tanaka, Y. Hozumi, S. Endo, T. Kudo, and K. Taniguchi, *Kobunshi Kagaku* 20, 687 (1963); *Chem. Abstr.* 60, 14615c (1964).

Small-Scale Production of Macroreticular Polystyrene Beads

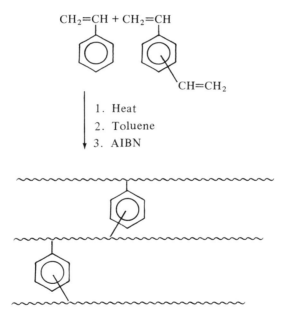

Submitted by D. C. Sherrington (1)

Checked by J. A. Moore and J. J. Kennedy (2)

1. Procedure

Macroreticular resins are highly cross-linked polymer beads in which large pore volumes are generated by the inclusion of a suitable inert solvent during suspension polymerization (3,4). Such resins have proved extremely valuable in the development of gel permeation chromatography (5). During their preparation the efficiency of maintaining suspension is crucial for success, particularly in small-scale production (10 g) where the volume/surface-area ratio is small, and therefore the tendency towards aggregation is exaggerated. The experimental technique reported here is based on general principles described in detail elsewhere (6) and allows preparations to be carried out in a highly reproducible manner.

A 250-ml round-bottomed flask with a B.34 Quickfit (equivalent to ℥ 34 ground glass joint) neck is modified by the introduction of four vertical baffles into its sides (Note 1). This modification is most conveniently achieved by appropriate melting and pressing of the glass. The baffles (B) should each protrude inwardly about one-tenth the diameter of the flask (see the modified flask and stirrer guide diagram below). The stabilizer, polyvinylpyrrolidone [0.5 g (M.W. ~24,000)], is weighed into the flask and dissolved in 100 ml of distilled water. A mixture of 7.5 ml of divinylbenzene and 2.5 ml of styrene (Notes 2 and 3) is then added, followed by 20 ml of toluene, the diluent (Note 4). Finally 0.1 g of the free radical initiator, azobisisobutyronitrile, is introduced and the mixture is rapidly agitated to ensure dissolution of the various components.

The flask is clamped in a thermostated bath and the stirrer arrangement is connected. The glass stirrer is the propeller type with an impeller diameter of about one-third to one-quarter the diameter of the flask (hence the large neck). The stirrer is positioned so that the impeller sits about two-thirds of the distance below the surface of the mixture (Note 5). The guide can be the conventional mercury seal type, but more regular and reproducible rotation is achieved by the use of a guide fitted with Teflon® bushes (T). In the latter modification a small condenser (C) may also be incorporated to prevent solution loss by evaporation. A constant torque motor is used to drive the stirrer at 400-500 rpm, the direction of rotation being such that the impeller drives the suspension *towards the bottom of the flask*.

The whole apparatus is flushed with nitrogen gas and the bath temperature is adjusted to 80°. Polymerization is allowed to proceed for 6 hr. After this time lapse the stirrer is stopped and the mixture is allowed to cool. The polymer beads, swollen with toluene, are then collected by suction filtration and washed twice with tetrahydrofuran. All solvents are removed by drying at 40° overnight in a vacuum oven. Traces of stabilizer are readily removed by Soxhlet extraction with tetrahydrofuran for 6 hr (Note 6). The yield is virtually quantitative.

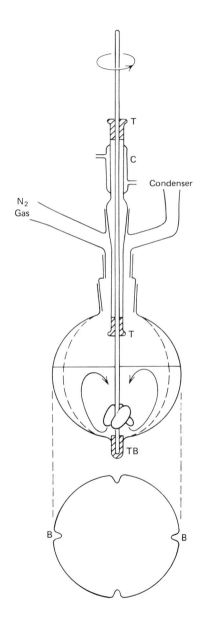

T

C

Condenser

N$_2$
Gas

T

TB

B   B

2. Characterization

The dry beads have spherical symmetry and a distribution of sizes in the range 30-100 mesh (Note 7). Particle sizes less than 100 mesh can be achieved by varying the experimental parameters (e.g., stirrer speed, weight of stabilizer, etc., though the use of inorganic stabilizers may also become necessary) with the accompanying problem of separation and purification. Beads larger than 30 mesh are difficult to prepare reproducibly on a small scale. Porosity can be reduced by limiting the volume of diluent; generally, beads will absorb slightly larger volumes of solvent than that used in their preparation. Cross-link density (Note 3) can also be varied readily, though if only small quantities of difunctional monomer are to be used the amount of diluent must simultaneously be restricted. Very low cross-link ratios may also require the use of inorganic stabilizers.

3. Notes

1. The optimum design for a baffled reaction vessel is one with a cylindrical geometry and flat base. The checkers used a standard Morton flask fitted with a U-tube adapter and a stirrer with a ground-glass bearing.

2. The commercial-grade monomers are stabilized with 0.1% *tert*-butylcatechol, which is readily removed by washing with dilute caustic soda solution and twice with water.

3. Divinylbenzene is usually supplied as a 50-60% w/w mixture with ethylvinylbenzene. This preparation therefore uses ~37% difunctional monomer giving a cross-link ratio of similar value.

4. Toluene is a good solvent for both the monomers and the polymer. The use of a diluent in which only the monomers are soluble, for example, aliphatic alcohols, produces cross-link networks which are even more porous (3). The morphology of these resins is complicated and certainly differs from those produced using diluents which readily swell the polymers.

5. More reproducible setting of the impeller height can be achieved using a flask with a Teflon bushing (TB) set in its base. The stirrer shaft can then be extended to seat in the bushing and provide a fixed impeller position.

6. The beads readily take up large volumes of both solvents and nonsolvents, swelling by a factor of about 3 in tetrahydrofuran. Soxhlet thimbles should be filled accordingly.

7. The checkers did not characterize the size distribution of their sample.

4. Methods of Preparation

Macroreticular resins are highly porous and are finding increasing use in solid-phase reactions as pioneered by Merrifield (7). The present preparation has the advantage of being reproducible and also yielding a product which is very easily isolated and purified. Functionalized resins are readily prepared in a similar way, by incorporation of the required amounts of an appropriately substituted comonomer. The simple purification procedure ensures that such functional groups remain intact for subsequent exploitation.

5. References

1. Department of Pure and Applied Chemistry, University of Strathclyde, 295 Cathedral St., Glasgow G1 1XL, Scotland.
2. Department of Chemistry, Rensselaer Polytechnic Institute, Troy, New York 12181.
3. J. C. Moore, *J. Polymer Sci.* A2, 835 (1964).
4. R. Kunin, E. Meitzner, and N. Bortnick, *J. Amer. Chem. Soc.* 80, 305 (1962).
5. W. Heitz, *Angew. Chem. Int. Ed.* 9, 689 (1970).
6. J. H. Rushton, Chapter IV in *Techniques of Organic Chemistry*, Vol. III, Part II, A. Weissberger, Ed., Interscience, New York, 1957, p. 235.
7. R. B. Merrifield, *Biochemistry* 3, 1385 (1964).

Acknowledgment

The author wishes to thank Mr. A. Hunter for experimental assistance.

Phenylated Polyimides

{Poly[Oxy-6,2-(1,3-Dioxo-4,5,7-Triphenyl-isoindolinediyl)-1,4-Phenyleneoxy-1,4-Phenylene-2,6-(1,3-Dioxo-4,5,7-Triphenylisoindolinediyl)-1,4-Phenylene]}

Submitted by F. W. Harris and W. A. Feld (1)
Checked by J. A. Moore and P. F. Holmes (2)

1. Procedure

A. Preparation of Dianhydride

A mixture of 11.42 g (12.4 mmole) of 3,3'-(oxydi-*p*-phenylene)-*bis*(2,4,5-triphenylcyclopentadienone) (3), 2.44 g (25.0 mmole) of maleic anhydride, and 30 ml of bromobenzene is placed in a 100-ml round-bottomed flask and heated at reflux for 3 hr. The reaction mixture is allowed to cool, and a solution of 2.2 ml of bromine in 3.5 ml of bromobenzene is added slowly (4). After the exothermic reaction has subsided, the mixture is heated at reflux for 3 hr. The solution is allowed to cool and then slowly added to 500 ml of petroleum ether (b.p. 60-110°). The precipitate is collected, recrystallized twice from toluene, and dried *in vacuo* at 153° for 24 hr to yield 10.11 g (75%) of white product [m.p. 250-255°, 330-332° (DSC) (Note 1); ir (cm^{-1}) 1840, 1780 (anhydride) (5,6)].

B. Polymerization

In a dry, 25-ml, three-necked flask equipped with a magnetic stirring bar, a nitrogen inlet, a short-path distillation apparatus, and a stopper are placed 0.1089 g (0.544 mmole) of 4,4'-diaminodiphenyl ether (Note 2), 3 ml of *m*-cresol (Note 3), and 0.05 g of isoquinoline. After the diamine is dissolved (Note 4), 0.5000 g (0.544 mmole) of the dianhydride is added in several portions over a period of 1 hr. The final portion of dianhydride is dissolved in 1 ml of *m*-cresol, and the solution is added to the reaction mixture. The mixture is then stirred for 1 hr at room temperature to yield a viscous, bright yellow, polyamic acid solution (Note 5). The stopper is replaced by an addition funnel, and the temperature of the flask is increased until distillation commences (250-280°). The volume of the solution is maintained at approximately 4 ml by continually replacing the distillate with *m*-cresol containing 0.025 g/ml of isoquinoline (Note 6). The distillation-addition cycle is carried out for 3 hr. After the viscous solution is allowed to cool, it is diluted with 15 ml of chloroform and slowly added to 300 ml of vigorously stirred absolute ethanol. The polymer that precipitates is reprecipitated from chloroform with absolute ethanol, collected, and air-dried to yield 0.520 g (94%) of product. The white polymer is then heat-treated to insure complete conversion of the intermediate polyamic acid to the polyimide (Note 7). A film of the polymer is cast from a 5% chloroform solution on the bottom of a 50-ml round-bottomed flask by slowly removing the solvent under reduced pressure. The film is heated under nitrogen at 275-300° for 4 hr. The heat-treated polymer is precipitated from chloroform with absolute ethanol, collected, and dried *in vacuo* at 153° for 48 hr (Note 8).

2. Characterization

The polymer is soluble in *m*-cresol, chloroform, and *sym*-tetrachloroethane. The intrinsic viscosity determined in *sym*-tetrachloroethane ranges from 1.4 to 2.8 dl/g (Note 8). Films can be cast from chloroform which are transparent and highly flexible. The ir spectra of the films show bands at 1780 and 1730 cm⁻¹, characteristic of imides. Differential scanning calorimetry and softening-under-load (11.2 psi) measurements indicate a T_g for this polymer at 360°. Thermogravimetric analysis (5°/min) shows that the onset of thermal decomposition of the polyimide occurs at ~530° in both nitrogen and air.

3. Notes

1. The melting point at 250-255° is followed by an exothermic crystallization.

2. The following diamines have been used: *p*-phenylenediamine, *m*-phenylenediamine, 4,4'-diaminodiphenyl ether, 4,4'-diaminodiphenylmethane, 4,4'-diaminobiphenyl, and 1,3-di-(3-aminophenoxy)benzene (6). All diamines were sublimed twice prior to use except 4,4'-diaminobiphenyl, which was recrystallized from absolute ethanol. Table 1 lists representative diamines and methods of purification.

3. The freshly distilled *m*-cresol was deoxygenated at 100° by bubbling nitrogen through the heated solution.

4. All the diamines except 4,4'-diaminobiphenyl dissolved quickly. Complete solution is not necessary for the reaction to proceed.

Table 1
Purification of Diamine Monomers

Diamine	Purification[a]	Melting Point (°)
1,4-Phenylene diamine	A	139-140
1,3-Phenylene diamine	B	61-63
4,4'-Diaminodiphenyl ether	C	189-191
4,4'-Diaminodiphenylmethane	D	91-93
4,4'-Diaminobiphenyl	E	127-129
1,3-Di-(3-aminophenoxy)benzene	F	114-116

[a]Purification conditions: A. Sublimed twice at atmospheric pressure. B. Sublimed twice at 90-100°/1 torr. C. Sublimed twice at 170-190°/1 torr. D. Sublimed twice at 110-120°/1 torr. E. Recrystallized twice from absolute ethanol and dried under vacuum. F. Sublimed twice at 120-140°/1 torr (sublimes as an oil which crystallizes on cooling.).

5. The solution may be opaque until heat is applied.

6. About 40 ml of this solution is needed and should be mixed immediately prior to use.

7. It estimated from ir analysis that 1-5% of the imide rings are not closed at this point.

8. The checkers obtained an intrinsic viscosity of 0.7.

4. References

1. Department of Chemistry, Wright State University, Dayton, Ohio 45431.
2. Department of Chemistry, Rensselaer Polytechnic Institute, Troy, New York 12181.
3. M. A. Ogliaruso, L. A. Shadoff, and E. I. Becker, *J. Org. Chem.* **28**, 2725 (1963).
4. O. Grummitt, in E. C. Horning (editor), *Organic Syntheses, Collective Volume 3*, p. 807.
5. F. W. Harris, S. O. Norris, L. H. Lanier, and W. A. Feld, *Am. Chem. Soc. Div. Org. Coat. Plast. Preprints*, **33**(1), 160 (1973).
6. F. W. Harris and W. A. Feld, unpublished results.

Acknowledgment

We wish to acknowledge the support of the Research Corporation, which helped make this synthesis possible.

Free-Radical Polymerization of 4-Vinyl Pyridine; N-Alkylation of the Polymer

{Poly[1-Butyl-4-Pyridinio)ethylene Bromide]}

Submitted by D. Ghesquiere, J. Morcellet-Sauvage, and C. Loucheux (1)

Checked by J. A. Moore and J. J. Kennedy (2)

1. Procedure

A. Polymerization of 4-Vinylpyridine (4VP)

A solution containing 20 ml of freshly distilled 4VP (Note 1), 40 ml of methanol (Note 2), and 0.02 g of α,α'-azoisobutyronitrile (Note 3) is introduced into a polymerization tube which can be sealed. This vessel is connected to a high vacuum line (10^{-5} torr). The monomer solution is frozen and evacuated four times successively. The vessel is sealed to eliminate the influence of dissolved gas, especially oxygen, during the polymerization. After sealing, the polymerization vessel is maintained for at least 6 hr at 60° in a thermostated bath. The solution becomes progressively viscous. At the end of the polymerization reaction a slight pink color may appear because of oxidation of the monomer, even under the careful conditions used.

The crude polymer is recovered by precipitation in ethyl ether. The polymer obtained is white. Further precipitations may be performed from methanol solutions, using dry ethyl ether as a nonsolvent. The polymer is dried under vacuum at 50°.

B. Alkylation of Poly(4-Vinylpyridine)

A convenient solvent for alkylation is tetramethylene sulfone (TMS) (Note 4). A standard procedure is given for alkylation with butyl bromide (Note 5). P4VP (5 g) is added to 100 ml of TMS at 40° and the mixture is stirred until complete dissolution (Note 6). Butyl bromide (25 g) is then added to the P4VP solution and the reaction vessel is maintained at 70° during the time of the reaction (Note 7). The reaction is carried out in a vessel equipped with a ground glass stopper, in the presence of air or under an oxygen-free nitrogen flow (Note 8). Magnetic stirring is maintained throughout the reaction.

The quaternized polymer is recovered by precipitation in dry ethyl ether. Because of its hygroscopic behavior the quaternized polymer must be stored in the absence of moisture.

2. Characterization

A. Poly(4-Vinylpyridine)

The molecular weight of poly(4-vinylpyridine) (P4VP) is determined from measurement of its intrinsic viscosity in ethanol at 25°, using a viscosity/molecular-weight relationship given by Berkowitz (3):

$$[\eta] = 25.0 \times 10^{-5} \, \bar{M}_w^{0.68}$$

B. Quaternized P4VP

During the alkylation reaction (or at its end) the percentage of alkylated groups can be determined by potentiometric titration of Br⁻ ions with 0.05 N AgNO$_3$ using a silver electrode and a glass reference electrode. An aliquot is taken and added to 25 ml of methanol and 2.5 ml of 2 N H$_2$SO$_4$ (to stabilize the glass electrode used as a reference). The addition of methanol stops the alkylation if the titration is carried out during the course of the alkylation reaction.

3. Notes

1. Purification of 4VP (a Fluka "purum" product). After washing with 0.1 N NaOH solution to remove the inhibitor and drying over calcium hydride, monomeric 4VP is distilled three times under reduced pressure in the presence of pulverized calcium hydride.

2. Methanol is a reagent grade product. Chloroform can be used alternatively as a solvent for the polymerization reaction.

3. α,α'-Azoisobutyronitrile is a Fluka "purum" product. The checkers found it best to recrystallize the initiator from absolute methanol.

4. Tetramethylene sulfone (Merck, m.p. 27°) is distilled under reduced pressure before use.

5. Butyl bromide is a Fluka "purissimum" product. It can be used without further purification. However, if the bromide is slightly colored, distillation at 15 torr is required. Alkylation with other alkyl bromides such as hexyl or octyl bromide is quite similar.

6. The time needed for complete dissolution of the polymer can be as long as a few hours or more (depending on the molecular weight of the polymer).

7. When the percentage of pyridine groups quaternized is higher than 50%, the velocity of the reaction decreases strongly (the second-order velocity constant is reduced by one-tenth) and the time needed for complete alkylation increases dramatically (3-8). At 70°, the alkylation with butyl bromide reaches a maximum after three days. If hexadecyl bromide is used, a week is necessary.

8. In the presence of air, at 70°, the solution in TMS becomes yellow because of colloidal sulfur originating from thermal decomposition and/or oxydation of TMS. This slight decomposition does not influence the properties of the alkylated polymer, which is white after precipitation.

4. References

1. Université des Sciences et Techniques de Lille, B.P. 36, 59650 Villeneuve d'Ascq, France.
2. Department of Chemistry, Rensselaer Polytechnic Institute, Troy, New York 12181.
3. J. B. Berkowitz, M. Yamin, and R. M. Fuoss, *J. Polymer Sci.* **28**, 69 (1958).
4. R. M. Fuoss and B. D. Coleman, *J. Amer. Chem. Soc.* **77**, 5472 (1955).
5. R. M. Fuoss, M. Watanabe, and B. D. Coleman, *J. Polymer Sci.* **48**, 5 (1960).
6. C. B. Arends, *J. Chem. Phys.* **39**, 1903 (1963).
7. E. Tsuchida and S. Irie, *J. Polymer Sci.* **11**, 789 (1973).
8. J. Morcellet-Sauvage and C. Loucheux, *Makromol. Chem.* **176**, 315 (1975).

Poly[bis (m-Chlorophenoxy) phosphazene]

{Poly[Nitrilo(bis(3-Chlorophenoxy)-phosphoranylidyne)]}

Submitted by R. E. Singler, B. L. LaLiberte, and R. W. Matton (1)
Checked by S. Lund-Couchman, M. Tirrell, and O. Vogl (2)

1. Procedure

Hexachlorocyclotriphosphazene (*Caution!* Note 1) (50 g) is sealed in a Pyrex® tube (28 mm × 5 in.) (Note 2) which has been twice flushed with argon or nitrogen and evacuated to less than 0.1 torr. The polymerization is conducted in a thermoregulated heating block at 245° for ~30 hr. Initially the contents of the tube are rather fluid, but as the polymerization proceeds the reaction mixture becomes increasingly more viscous. The polymerization is terminated by removing the tube from the heating block and observing when the contents of the tube barely flow. Terminating the reaction at this point prevents extensive cross-linking and subsequent gelation, as indicated by the observation that the products are soluble in 100-150 ml of benzene (*Caution! Benzene is toxic and should be used under a hood.*) (Note 3). Most of the unreacted trimer, low-molecular-weight cyclic molecules, and oligomers are then separated from the high polymer by coagulation of the polymer with 350 ml of *n*-pentane. Yield of the polymer is 11.3 g (0.098 mole, 24%) (Note 4). The isolated polymer is immediately dissolved in 150 ml of toluene (Note 5).

A solution containing 30 g (0.23 mole) of *m*-chlorophenol (reagent grade, vacuum distilled), 100 ml of benzene, and 250 ml of *bis*(2-methoxyethyl)ether is dried by removal of ∼30 ml of benzene using a Dean-Stark trap. After cooling to room temperature, 5.15 g (0.22 mole) of sodium particles are slowly added under argon and allowed to react for several hours while raising the temperature to about 80°. Some precipitation and slight discoloration, which is probably caused by the formation of the aryloxide salt, occurs. (An alternative method for the preparation of sodium *m*-chlorophenoxide is to first prepare sodium methoxide with 5.15 g of sodium metal and an excess of methanol in 100 ml of diglyme. A solution containing 30 g of *m*-chlorophenol, 150 ml of diglyme and 200 ml of benzene is subsequently added dropwise with stirring. Upon heating, methanol is azeotropically removed with benzene. The temperature is maintained below 125° by the addition of more benzene as necessary. The methanol content is monitored routinely by gas chromatography using an 8 ft × ¼ in. stainless steel column packed with 10% SE30 on Chromosorb W, He flow 1 ml/min, at 100°.) When essentially all of the sodium has been consumed (Note 6) the polymer solution is added over a 60-min period to the aryloxide solution at 80°, resulting in an off-white emulsion. The temperature is slowly raised to 125° and maintained for 24 hr (Note 7) under a blanket of argon or nitrogen. The reaction is subsequently cooled and added dropwise to several liters of methanol to precipitate the polymer. The polymer is washed with water and methanol to remove sodium chloride and excess *m*-chlorophenol. The polymer is twice dissolved in tetrahydrofuran, filtered, and precipitated into several liters of methanol. After a final wash with methanol, the polymer is dried under vacuum (25°, 0.1 torr) to give 15 g of poly[*bis*(*m*-chlorophenoxy)phosphazene] (56% based on 11.3 g of poly(dichlorophosphazene). This procedure is generally applicable for other poly(aryloxyphosphazenes) (3,4).

2. Characterization

Analysis. Calcd. for ($C_{12}H_8O_2Cl_2NP$): C, 48.03%; H, 2.69%, Cl, 23.63%. Found: C, 48.07%; H, 2.74%; Cl, 23.17. Low-carbon and high-chlorine analyses would indicate incomplete substitution and presence of residual chlorine, respectively.

Poly[*bis*(*m*-chlorophenoxy)phosphazene] is soluble in tetrahydrofuran but insoluble in hexane, methanol, and acetone. Provided that precautions are taken during synthesis to avoid extensive branching and hydrolysis, the polymer is also soluble in chloroform, *sym*-tetrachloroethane, cyclohexanone, and benzene; otherwise, aggregation and swelling will occur in nonpolar solvents.

The ir spectrum (Note 8) showed no evidence for the presence of OH or POP units. The principal ir assignments are as follows (in cm^{-1}): 1590 (aromatic C—C), 1270 (P=N), 1210, 955 (P—O—C), 780 (P—N). The ^{1}H-nmr spectrum showed a broad multiplet at 6.7 δ assigned to aromatic protons, which did not resolve appreciably for temperatures up to 125°.

Thin, opaque films cast from tetrahydrofuran can be cold-drawn and are crystalline by x-ray analysis. Films routinely give elongations of several hundred percent and ultimate breaking strengths of 4000 to 8000 psi based on final cross-sectional area. However, compression-molded films (130°, 100 atm) are more brittle and give lower elongations and breaking strengths.

The glass transition temperature is −24°. A well-defined endothermic transition, designated T(1) (5), occurs in the region of 60-100°. Both the temperature and heat capacity change are dependent on thermal history. Slow cooling through the T(1) region increased the size of the transition and raises the temperature (Fig. 1). These measurements were made on a solution-cast film using a DuPont 990 thermal analyzer. The decomposition temperature (TGA) is 400° with a 50% weight loss at 450°.

The intrinsic viscosity and light scattering data measured in chloroform at 25° were: [η] = 1.66 dl/g, $\bar{M}_w$ = 6.13 × 10^6, $\langle S^2 \rangle_z^{\frac{1}{2}}$ = 1280 Å, and A_2 = 5.3 × 10^{-5}. Although the number average molecular weight $\bar{M}_n$ was too large to be accurately determined ($\bar{M}_n \geqslant 5 \times 10^5$), the polydispersity ($M_w/M_n$) was approximately 10. Preliminary indications based on light scattering data are that this polymer is not highly branched. This polymer and other poly(aryloxyphosphazenes) prepared by this reaction sequence (3) generally have a bimodal molecular weight distribution as shown by gel permeation chromatography (Note 9).

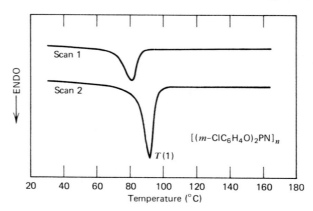

Fig. 1. Differential scanning calorimetry traces of a solution-cast film of poly[*bis*(*m*-chlorophenoxy)phosphazene]. The second scan shows the effect of slow cooling (160-30°) on the appearance of the $T(1)$ endotherm.

3. Notes

1. Caution should be exercised in the handling of hexachlorocyclotriphosphazene. It should be sealed when not in use, protective clothing should be worn, and it should be handled in a well-ventilated area. Commercial-grade hexachlorocyclotriphosphazene (Millmaster, Ethyl Corp.) is a mixture of cyclic molecules (trimer, tetramer, etc.) and oligomers. It can be prepared by known procedures (6). It was purified by recrystallization from hot heptane followed by vacuum distillation (130-140°/20 torr): softening point, ~90°; melting point 110-112°. Analysis by gas chromatography using an SE-30 column at 240° showed only trimer (≥98%) and tetramer (≤2%). Analysis by differential scanning calorimetry showed a slight endotherm at ~90° with the major melting transition at 114°. These data probably reflect the melting of the trimer-tetramer eutectic composition at 90° (7).

2. Constricted Pyrex® tubes were annealed, rinsed with distilled water and methanol, and dried at 110° prior to use.

3. Open-chain poly(dichlorophosphazene) is soluble in benzene, toluene, tetrahydrofuran, carbon tetrachloride, and diglyme. It is insoluble in pentane and most other aliphatic hydrocarbons. Reagent-grade benzene and toluene were dried azeotropically prior to use. *Bis*(2-methoxyethyl)ether (diglyme) was obtained from Ansul and used as received. The checkers recommend that hourly checks for gelation be made after 12 hr of reaction time.

4. Yields of soluble polymer may vary from 15 to 70% depending on the

purity of starting material, polymerization times and temperatures, polymer isolation procedures, and other factors (5). In general, polymer yields of 20-40%, obtained from about 30 hr of polymerization at 245°, insure the formation of soluble, open-chain poly(dichlorophosphazene). The yield of polymer (11.3 g) is the difference between the amount of reaction products initially dissolved in benzene and the amount of predominately cyclic molecules recovered from the benzene-pentane solution. The checkers obtained from twice-recrystallized (heptane) vacuum-distilled monomer a 10% yield after 55 hr. Twice-recrystallized monomer (decolorized with activated charcoal) with no distillation gave the checkers 40-70% yields.

5. To avoid hydrolysis and cross-linking, poly(dichlorophosphazene) is never dried, and it is used in the substitution reaction as soon as possible. However, either a benzene or toluene solution of poly(dichlorophosphazene) is reasonably stable for several days at room temperature.

6. To obtain the desired polymer, it does not appear necessary to react all of the sodium prior to the addition of poly(dichlorophosphazene). In general, it appears better to add the polymer to the aryloxide solution several hours after the sodium has reacted at 80°, rather than allowing the aryloxide mixture to react completely overnight or raising the temperature above 100°. Otherwise, lower molecular weights and lower yields of the desired polymer are often obtained. The checkers noted that reaction time for the sodium and *m*-chlorophenol reaction can be reduced to 1 hr if small (0.5-cm cubes) pieces of freshly cut sodium are used.

7. As the temperature is raised, solvent is removed. Use of the Dean-Stark trap permits effective control of the desired reaction temperature.

8. The ir spectrum was obtained on a vacuum-dried thin film cast on a sodium chloride plate and recorded with a Beckmann IR-12 instrument. The ^{1}H-nmr spectrum was obtained on a 2-wt.% solution in *sym*-tetrachloroethane-d_2 with hexamethyldisiloxane as a reference and recorded with a Perkin-Elmer R-32 90-MHz spectrometer.

4. References

1. Organic Materials Laboratory, Army Materials and Mechanics Research Center, Watertown, Massachusetts 02172.
2. Department of Polymer Science and Engineering, University of Massachusetts, Amherst, Massachusetts 01002.
3. R. E. Singler, G. L. Hagnauer, N. S. Schneider, B. R. LaLiberte, R. E. Sacher, and R. W. Matton, *J. Polym. Sci. Polym. Chem.* **12**, 433 (1974).
4. H. R. Allock, R. L. Kugel, and K. J. Valan, *Inorg. Chem.* **5**, 1709 (1966).

5. R. E. Singler, N. S. Schneider, and G. L. Hagnauer, *Polymer. Eng. Sci.* **15**, 321 (1975).
6. N. L. Nielsen and G. Cranford, *Inorg. Synthesis* **6**, 94 (1960).
7. L. F. Audrieth, R. Steinman, and A. D. F. Toy, *Chem. Rev.* **32** 109 (1943).

Acknowledgment

The authors acknowledge the assistance of Dr. Gary Hagnauer, Army Materials and Mechanics Research Center, for assistance in obtaining and interpreting the solution characterization data.

Alternating Cyclopolymerization of Maleic Anhydride and Divinyl Ether

{Poly[2,4-(7,9-Dioxo-3,8-Dioxabicyclo[4.3.0]-nonanediyl)-3,4-Dioxooxolanediyl)methylene]}

Submitted by G. B. Butler (1) and C. C. Wu (2)

Checked by S. C. Chu (3)

1. Procedure

A 100-ml round-bottomed three-necked flask (Note 1) equipped with (i) a stirrer with a ground-glass bearing and a half-moon Teflon® blade, (ii) a dry-ice condenser having a nitrogen inlet connected by vacuum tubing to a three-way stopcock which is connected to a purified N_2 (Note 2) source and a vacuum pump, and (iii) a septum, is evacuated while heating (Note 3) and then filled with pure, dry nitrogen. After this evacuation and nitrogen-filling procedure is repeated several times, the septum is removed while a positive pressure of nitrogen is passing through the flask. While it is immersed in an ice-water bath and is purged with a slow stream of N_2, the flask is charged, through the open neck, with 40 ml of purified benzene (*Caution!* Note 4), 1.962 g (20 mmole) of purified maleic anhydride (MA) (*Caution!* Note 5), and 13.6 mg (0.1 mmole) of purified α,α-azobisisobutyronitrile (AIBN) (Note 6). The dry-ice condenser is filled with dry ice, and then 1.0 ml (0.77 g, 11 mmole) of purified divinyl ether (DVE) (Note 7) is charged to the flask with a syringe. Thereafter, a thermometer is quickly attached to the open neck, the nitrogen flow is stopped, and stirring is started. The MA is completely dissolved in the benzene in about 1 min. Next, the ice-water bath is removed from the flask, and a silicone oil bath maintained at 70° is raised to heat the flask and to initiate the cyclocopolymerization. The clear solution becomes cloudy about 10 min after the start of heating, and the cloudiness intensifies as polymerization proceeds. There is little or no noticeable reflux of divinyl ether during the course of the cyclocopolymerization. Heating is continued for a total time of 100 min after which the flask is cooled to room temperature with an ice-water bath. The copolymer is then filtered with a sintered glass funnel and dried at 50° *in vacuo* overnight. The white copolymer is obtained in the yield of 2.267 g (Note 8). The copolymer is purified once by dissolving it in purified acetone (Note 9) and precipitating it from purified benzene (Note 4). It is again dried at 50° *in vacuo* overnight to give 2.094 g (78.6% conversion of the copolymer).

2. Characterization

The copolymer is soluble in polar solvents such as acetone, cyclohexanone, cyclopentanone, α-methyl-γ-butyrolactone, N,N'-dimethylformamide, and dimethylsulfoxide. It is partially soluble in methyl ethyl ketone and acetophenone, and is insoluble in aliphatic, aromatic, and chlorinated hydrocarbons, and ethers. The intrinsic viscosity in purified acetone (Note 9) at 30° is 0.75 dl/g.

Analysis. Calcd. for 2:1 ratio of maleic anhydride to divinyl ether copolymer: C, 54.13%; H, 3.75%; O, 42.11%. Found: C, 54.07%; H, 3.78%; O, 42.18%.

The ir spectrum of the copolymer exhibits the following absorption bands (cm^{-1}): 1860 and 1790 s (cyclic anhydrides); 1230 s (six-membered cyclic ether and alkanes skeletal vibrations); 1100-1080 m (six-membered cyclic ether). The copolymer may also include some five-membered cyclic ethers.

The 60-MHz nmr spectrum in acetone-d_6 at 50° shows rather broad multiplet peaks (δ) centered at 2.32 (H_1, H_2, H_6, and H_7), 3.47 (H_9, H_{10}), 4.00 (H_4, H_5), and 4.50 (H_3, H_8). (The peak at 2.32δ also exhibits a pronounced shoulder on the high field side.) having an area ratio of 2:1:1:1. The nmr spectrum of the copolymer prepared from deuterated maleic anhydride demonstrates the disappearance of the peaks at 3.47δ and 4.00δ, and a separation of the 2.32δ peak into two distinct multiplets centered at 2.30δ and 2.08δ, indicative of the $C_6 C_7$ and $C_1 C_2$ pairs. A gel permeation chromatographic study (4) indicates that molecular weight measurements of the copolymer by GPC can be made by chemically converting the acid anhydrides of the copolymer to dimethyl esters, then running the GPC fractionation of the derived copolymer with Styragels® in tetrahydrofuran, and making the molecular weight calculations with the aid of the polystyrene universal calibration.

The checker reported a glass transition temperature of 200° for the copolymer, as determined by differential scanning calorimetry.

3. Synthesis of Divinyl Ether (Note 7) (5)

A 3-liter three-necked flask is equipped with a mechanical stirrer, an addition funnel, a reflux condenser, and a thermometer. Into this flask is placed 1000 g (17.9 mole) of analytical-grade potassium hydroxide and 200 g (1.41 mole) of triethanolamine. With the object of removing divinyl ether from the sphere of reaction as fast as it is formed, warm water (35°) is passed through the reflux condenser, which is connected to a condenser with ice-water cooling. A two-necked round-bottomed flask as receiver is connected to this distillation condenser and immersed in an ice-water bath. The second neck of the receiver is connected with a drying tube. The mixture of potassium hydroxide and triethanolamine is preheated until the KOH melts (about 190-210°). With stirring, 400 g (2.79 mole) of *bis*(2-chloroethyl ether) [(*Caution!* Note 10) b.p. 91°/37 torr] is added slowly from the dropping funnel to the alkaline solution with the mixture temperature between 160 and 190°. White fumes are observed immediately after addition is begun. Liquid divinyl ether is collected for 2 hr after all starting material is added. Further refluxing does not improve the yield significantly. The product is washed three times with precooled water (to prevent evaporation of divinyl ether) followed by washing three times each with cold hydrochloric acid (5%) and cold deionized water. The resulting organic layer is dried over calcium chloride overnight. After refluxing and distilling over calcium hydride, the product is kept in the refrigerator. The purified product consists of 75 g (27% yield) of pure divinyl ether, b.p. 29-29.5° [(literature (5) b.p. 28-29°]. The ir spectrum is identical to the reported spectrum. The nmr spectrum shows a clear ABX pattern in the olefinic hydrogen region. Few or no impurities are detectable in the nmr spectrum.

4. Notes

1. The glassware used is soaked in a dichromate cleaning solution overnight, then thoroughly rinsed with tap water and subsequently with distilled water, followed by oven-drying prior to use.

2. Water in the commerically available prepurified nitrogen is removed by passing the nitrogen through a bubbler and a 30-cm-long glass tube filled with Linde 3A molecular sieves, which is immersed in a dry-ice-isopropanol cold bath.

3. A Bunsen burner, heat gun, or a commerical hair dryer may be used to heat the flask.

4. Benzene is a toxic substance and should be used in a hood. Analytical-grade benzene is stirred with concentrated sulfuric acid for two days. It is then washed with dilute aqueous KOH solution several times, followed by washing with water several times. The washed benzene is dried over Linde 3A molecular sieves and distilled over phosphorous pentoxide in a nitrogen atmosphere.

5. Maleic anhydride is a toxic substance. Avoid skin contact. Maleic anhydride, m.p. 56°, is purified prior to use by recrystallization from analytical-grade benzene, and is further purified by sublimation under reduced pressure. The purified maleic anhydride is quickly ground to small pieces in a mortar to facilitate weighing.

6. α,α-Azobisisobutyronitrile is crystallized from absolute methanol, filtered, and dried *in vacuo* in the presence of P_2O_5.

7. Divinyl ether, formerly available as Vinethene® from Merck, Sharp & Dohme, is purified by first washing three times with precooled dilute aqueous NaOH solution while being cooled in an ice-water bath to minimize divinyl ether evaporation. This procedure is followed by washing three times with pre-cooled distilled water under the same conditions. The monomer is then dried over calcium hydride overnight and distilled from calcium hydride (b.p. 28.3°) under nitrogen. The purified divinyl ether is capped with a rubber septum and kept in a refrigerator until used. At the time this procedure was checked, divinyl ether was no longer commercially available. It was synthesized by the published procedure (5). It is now commercially available from Fairfield Chemical Co.

8. The checker obtained 1.96 g.

9. Acetone should be free from water which may cause hydrolysis of the copolymer. It is dried over Linde 3A molecular sieves and distilled from P_2O_5.

10. *Bis*(2-chloroethyl ether) is a toxic substance and should be used in a hood. Gloves should be warn to avoid skin contact.

5. Method of Preparation

Free-radical cyclocopolymerization of maleic anhydride and divinyl ether has been reported in several papers (6) and patents (7). A study of cyclocopolymerization in a number of solvents of different dielectric constant was published recently (8). Thermal autopolymerization, photopolymerization, and γ-ray polymerization (all in bulk) of maleic anhydride with divinyl ether, leading to the same alternating copolymer structure, were also investigated (8).

Homogeneous cyclocopolymerization of maleic anhydride and divinyl ether with free-radical initiators can be accomplished in acetone (4, 8), cyclohexanone, and cyclopentanone, but usually does not give high-molecular-weight copolymer probably because of chain transfer to the solvents and relatively low charge-transfer complexation between maleic anhydride and divinyl ether in these solvents. Inverse solubility phenomena are observed at certain high concentrations of monomers and at or above certain polymerization temperatures. Heterogeneous radical-initiated cyclocopolymerizations occur in ethylacetate, acetophenone, nitromethane, in aromatic solvents such as benzene, toluene, and xylenes (4,7), and in chlorinated hydrocarbons (8) such as chloroform, methylene chloride, and ethylene chloride. Mainly because of the high tendency for the comonomers to form a charge-transfer complex in the above solvents, higher polymerization rates and higher molecular weights of the copolymer are realized in comparison with homogeneous cyclocopolymerization (8,9).

6. References

1. Center for Macromolecular Science and Department of Chemistry, University of Florida, Gainesville, Florida 32611.
2. Monsanto Triangle Park Development Center, P.O. Box 12274, Research Triangle Park, North Carolina 27709.
3. Research and Development Department, Hooker Research Center, Long Road, Grand Island, New York 14072.
4. G. B. Butler and C. C. Wu, in *Water Soluble Polymers*, N. M. Bikales, Ed., Plenum, New York, 1973, p. 369.
5. M. F. Shostakovskii and E. V. Dubrova, *Bull. Acad. Sci. U.S.S.R. Chem. Sci.*, 319 (1958) [English translation].
6. G. B. Butler, *J. Polym. Sci.* 48, 279 (1960); G. B. Butler, *J. Macromol. Sci. Chem.* A5(1), 219 (1971).

7. G. B. Butler, U.S. Patent 3,320,216 (May 16, 1967); U.S. Patent Reissue 26,407 (June 11, 1968).
8. G. B. Butler and K. Fujimori, *J. Macromol. Sci. Chem.* **A6**(8), 1533 (1972).
9. B. Zeegers and G. B. Butler, *J. Macromol. Sci. Chem.* **A6**(8), 1569 (1972).

Poly [N-Acryloyl-N, N-bis (2,2-Dimethoxyethyl)amine]

{Poly[bis(2,2-Dimethoxyethyl)carbamoyl-ethylene]}

$$NH_2-CH_2-CH+(OCH_3)_2 + Br-CH_2-CH+(OCH_3)_2 \xrightarrow[100°]{aq.\ Na_2CO_3}$$

$$HN+CH_2-CH+(OCH_3)_2]_2 \xrightarrow[0°]{CH_2=CHCOCl}$$

$$CH_2=CHCON+CH_2-CH+(OCH_3)_2]_2 \xrightarrow[40°]{(NH_4)_2S_2O_8}$$

$$\left[\begin{array}{l} CH_2-CH \\ \quad\ \ | \\ \quad\ \ C=O \\ \quad\ \ | \\ \quad\ \ N+CH_2-CH+(OCH_3)_2]_2 \end{array} \right]_n$$

Submitted by R. Epton, B. L. Hibbert, and G. Marr (1)
Checked by C. G. Overberger and M. D. Shalati (2)

1. Procedure

A. N,N-bis(2,2-Dimethoxyethyl)amine

A two-liter round-bottomed standard polymerization flask equipped with a mechanical stirrer (Note 1), a condenser, and a 500-ml dropping funnel is placed in an oil bath. The flask is charged with a slurry consisting of 212 g (2 moles) of sodium carbonate in a solution of 315 g (3 moles) of aminoacetaldehyde dimethylacetal (Note 2) in 500 ml of water, and the stirrer motor is started.

After allowing 20 min for the contents of the flask to reach 100°, the dropping funnel is charged with 338 g (2 moles) of bromoacetaldehyde dimethylacetal (Note 3). This compound is added to the mixture over a further 20 min. The two-phase system is maintained at 100° and stirred continuously for 48 hr.

After cooling the reaction flask and its contents to 0° with an external ice bath, the upper organic layer is removed by siphoning (Note 4). The aqueous layer is mechanically stirred at 0° and sodium chloride (300 g) is added followed by 500 ml of ether. After 15 min the ether is again removed. The aqueous slurry is extracted nine more times with 500-ml aliquots of ether in a similar manner. The original organic layer and ether extracts are combined and dried over $MgSO_4 \cdot \frac{1}{2}H_2O$ overnight before the ether is removed by rotary evaporation. A dark brown oil is obtained which is distilled under reduced pressure using a 25 × 2-cm vacuum-jacketed fractionating column packed with glass helices. The first fraction to distill is 194 g (39%) of *N,N-bis*(2,2-dimethoxyethyl)amine, b.p. 100-114°/5 torr, n_D^{25} = 1.426870. *Analysis.* Calcd. for $C_8H_{19}NO_4$: C, 49.73%; H, 9.91%; N, 7.25%. Found: C, 49.32%; H, 10.71%; N, 7.25%. The residual oil in the distillation flask is largely crude *N,N,N-tris*(2,2-dimethoxy-ethyl)amine (Note 5).

B. *N-Acryloyl-N,N-bis(2,2-Dimethoxyethyl)amine*

It is essential that this reaction is carried out in a fume hood. A 500-ml three-necked round-bottomed flask equipped with a mechanical stirrer, a calcium chloride tube, and a 250-ml dropping funnel is immersed in an ice bath. A mixture of acryloyl chloride [4.52 g (0.05 mole)] (*Caution!* Notes 2 and 6) and 100 ml of dry, peroxide-free ether is placed in the flask, the stirrer motor is started, and the temperature is allowed to fall to 0° over a period of 20 min. A mixture consisting of 19.3 g (0.10 mole) of *N,N-bis*(2,2-dimethoxyethyl)amine and 100 ml of dry, peroxide-free ether is placed in the dropping funnel and addition commenced at such a rate that the temperature does not rise above 5°. A white precipitate of *N,N-bis*(2,2-dimethoxyethyl)amine hydrochloride appears immediately and the reaction mixture soon becomes a white slurry. After the addition of *N,N-bis*(2,2-dimethoxyethyl)amine is complete, stirring is continued for a further 30 min. The slurry is then swirled rapidly and filtered on a 250-ml Büchner funnel mounted on a 1-liter filter flask. The reaction flask is washed with 50 ml of dry, peroxide-free ether at 0° and the washings are poured quickly over the amine hydrochloride. The washings are transferred to a 500-ml round-bottomed flask and the ether is removed very quickly at room temperature using a rotary evaporator. A mobile oil results containing a small amount of ether, which is removed by purging the oil with a gentle stream of dry air for 3 hr (Note 7) to give 11.9 g (97%) of pure *N*-acryloyl-*N,N-bis*(2,2-dimethoxy-ethyl)amine, b.p. 120-123°/1 torr, n_D^{25} = 1.46061. *Analysis.* Calcd. for $C_{11}H_{21}NO_5$: C, 53.49%; H, 8.57%; N, 5.67%. Found: C, 53.25%; H, 8.66%; N, 5.60%.

C. Homopolymerization of N-Acryloyl-N,N-bis(2,2-Dimethoxyethyl)amine

Three gas wash bottles A, B, and C are arranged in series. Bottles A and B each contain 200 ml of 90% ethanol-water (Note 8) to saturate the stream of nitrogen before it enters bottle C, which serves as the polymerization vessel. Bottle C is charged with a mixture of 4.94 g (0.02 mole) of *N*-acryloyl-*N,N-bis*(2,2-dimethoxyethyl)amine and 9.0 ml of 90% ethanol-water, and a slow stream of nitrogen is passed through the mixture. After 1 hr, homopolymerization is initiated by adding 1.0 ml of a 5% w/v solution of ammonium persulfate in oxygen-free 90% ethanol-water. The gas inlet is arranged so that nitrogen blows over the surface of the liquid, and the reaction bottle is maintained at 40° ± 0.01° for 96 hr. The homopolymer is recovered from solution by diluting with 90% ethanol (5 ml) and pouring the liquid into distilled water (50 ml). A soft, white, sticky solid is immediately obtained, which is then left to harden in the water for several hours.

2. Characterization

The homopolymer is insoluble in water, ether, methanol, and petroleum ether, but is soluble in ethanol, chloroform, carbon tetrachloride, acetone, and benzene. The ir spectrum (film) gave peaks at 1645 (amide C=O stretch), 1130 and 1090 cm^{-1} (acetal C−O stretch). The intrinsic viscosity of the homopolymer in chloroform at 25° is 0.575 dl/g. *Analysis.* Calcd. for $C_{11}H_{21}NO_5$: C, 53.49%; H, 8.57%; N, 5.67%. Found: C, 53.47%; H, 8.77%; N, 5.53%.

3. Notes

1. A sturdy, crescent-shaped stirrer blade is essential to obtain efficient stirring of the thick reaction slurry.

2. Aminoacetaldehyde dimethylacetal and acryloyl choride were obtained from the Aldrich Chemical Co. and were sufficiently pure to use as received.

3. Bromoacetaldehyde dimethylacetal was prepared according to the method of Bedoukian (3).

4. This procedure is facilitated by the use of a wide-necked polymerization flask, the top of which may be removed during the extraction.

5. The yield of *N,N,N-tris*(2,2,2-dimethoxyethyl)amine was 88.1 g (19%), b.p. 130-132° at 5 torr, n_D^{25} = 1.42886. *Analysis.* Calc. for $C_{12}H_{27}NO_6$: C, 51.23%; H, 9.68%; N, 4.98%. Found: C, 50.99%; H, 9.73%; N, 5.16%.

6. Acryloyl chloride is extremely lachrymatory.

7. Dry air is effective both in inhibiting polymerization and in removing the last traces of ether. Removal of ether under reduced pressure leads to premature polymerization.

8. It was necessary to use 90% (v/v) ethanol-water as solvent to insure complete dissolution of the initiator, ammonium persulfate.

4. References

1. Department of Physical Sciences, Wolverhampton Polytechnic, Wolverhampton WV1 1LY, U.K.
2. Department of Chemistry, University of Michigan, Ann Arbor, Michigan 48109.
3. P. Z. Bedoukian, *J. Am. Chem. Soc.* **66**, 651 (1944).
4. R. Epton, B. L. Hibbert, and G. Marr, *Polymer* **16**, 314 (1975).

(Poly(L-Lactide)

{Poly[L-Oxycarbonylethylidene]}

Submitted by O. Aydin and R. C. Schulz (1)

Checked by C. G. Overberger and D. Wolf (2)

1. Procedure (Note 1)

Pure L(-)-lactide (5 g, Note 2) and 50 ml of dry toluene (reagent grade) were combined with a magnetic stirring bar in a 100-ml reaction vessel (see diagram; a heavy-walled pressure tube, for example, Ace Glass, Inc. Catalogue No. 8651-10 can also be used). Nitrogen is introduced and the sealed reaction vessel is immersed in an oil bath maintained at $100 \pm 5°$ with venting and agitation.

Then, 0.5 ml of initiator solution (Note 3) is injected through the serum cap with a hypodermic syringe. After about 3-4 hr the reaction mixture is cooled and slowly added to 300 ml of stirred methanol. The precipitated polymer is redissolved in a small volume of chloroform and precipitated by dropping into hexane. It is filtered and dried *in vacuo* at $70°$ overnight. The polymer yield is 4.0-4.6 g (80-92% of theoretical yield).

2. Characterization

The poly-L-lactide prepared in this way is soluble in chloroform, methylene chloride, acetonitrile, hexafluoroisopropanol, trifluoroethanol, trifluoroacetic

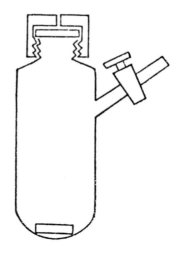

acid, or dioxane. It is insoluble in methanol, ether, and hexane. The melting point (153°) was determined by differential thermal analysis. Viscosity measurements were made in chloroform at 25° with an Ubbelohde viscometer (e.g., Cannon 50 L 813; η_{spec}/c was between 0.05 and 0.19 dl/g depending on the purity and water content of the polymerization mixture). The nmr spectrum in $CDCl_3$ contained resonances at δ = 1.6 (d, 3H) and 5.25 ppm (q, 1H) (4,7). The ir spectrum was obtained in KBr or as a film cast from hexafluoroisopropanol and revealed characteristic bands at 1755, 1090, 1130, 1180, and 1215 cm^{-1} (4). The optical dispersion and the circular dichroism spectra are reported in the literature (4,5,8); see Table 1. The conformation of the polymer is discussed in references 9 and 10.

Table 1
Rotation (Dioxane, c = 1 g/dl, 22°)

	Wavelength, λ (nm)			
	365	436	546	578
Monomer	-878°	-528°	-299°	-260°
Polymer	-428°	-295°	-182°	-161°

3. Notes

1. The polycondensation of lactic acid or its derivatives leads only to oligomers. High-molecular-weight polyesters of lactic acid are obtained only by ring-opening polymerization of lactide (3). If optically active lactide is used, an isotactic, optically active polylactide is formed (4-6). The configuration of the asymmetric carbon atom of the monomer is retained on polymerization with, for example, $SnCl_4$ or $ZnEt_2$.

2. Commercially available L(-)-lactide (M.W. 144) (e.g., Polyscience, Inc. Catalog No. 5749) may contain small amounts of water or free lactic acid. It is hygroscopic and must be stirred over KOH. Before use the monomer should be carefully recrystallized from anhydrous diethyl ether. Melting points: L-lactide (Et_2O), 95°; (D, L)-lactide (EtOH), 124.5°.

3. The initiator solution was 1.66 g of anhydrous stannic chloride in 5.0 ml of dried toluene. The solution must be clear and colorless.

4. References

1. Institute of Organic Chemistry, University of Mainz, D 65 Mainz, West Germany.
2. Department of Chemistry, University of Michigan, Ann Arbor, Michigan 48109.
3. J. Kleine and H. H. Kleine, *Makromol. Chem.* **30**, 23 (1959); T. Tsuruta, K. Matsuura, and S. Inoue, *Makromol. Chem.* **75**, 211 (1964).
4. R. C. Schulz and J. Schwaab, *Makromol. Chem.* **87**, 90 (1965); R. C. Schulz and A. Guthmann, *Polymer Lett.* **5**, 1099 (1967).
5. M. Goodman and M. D'Alagni, *Polymer Lett.* **5**, 515 (1967).
6. W. Dittrich and R. C. Schulz, *Angew. Makromol. Chem.* **15**, 109 (1971).
7. E. Lillie and R. C. Schulz, *Makromol. Chem.* **176**, 1901 (1975).
8. R. C. Schulz, IUPAC International Symposium on Macromolecular Chemistry, Budapest, 1969, pp. 185, 193.
9. P. DeSantis and A. J. Kovacs, *Biopolymers* **6**, 299 (1968).
10. A. E. Tonelli and P. J. Flory, *Macromolecules* **2**, 225 (1969).

Index